THE QUICK AND THE DEAD

THE QUICK AND THE DEAD

by
W.A. Waterton

Grub Street • London

First published 1956 by Frederick Muller Ltd

This edition published in 2012 by
Grub Street
4 Rainham Close
London SW11 6SS

British Library Cataloguing in Publication Data

Waterton, William Arthur.
  The quick and the dead.
  1. Airplanes--Great Britain--Flight testing--History.
  I. Title
  629.1'3453'0941'09045-dc23

  ISBN-13: 9781908117274

Cover design and formatting by Sarah Driver
Formatted by Sarah Driver

Printed and bound in Great Britain by MPG Ltd, Bodmin, Cornwall
Grub Street Publishing only uses
FSC (Forest Stewardship Council) paper for its books.

CONTENTS

DEDICATION

To Marjorie, who hounded me to write this book and kept my nose to the grindstone; who put up with my late sessions, kept me awake with coffee and provided the inspiration.

To "Spider" and "Bill", who, were they here, would have appreciated and understood.

To those few who design, build, operate and fly aircraft, and put ideals, practice and common-sense above profit and self-advancement.

To the many unheard of and unhonoured pilots who have devoted—and given—their lives to the air and the aeroplanes they loved.

To those frustrated designers, aerodynamacists, engineers and test pilots who have laboured honourably for their beliefs.

FOREWORD

I was delighted to be invited to write a foreword to this splendid book by Bill Waterton since my own experiences relate in so many ways to his, though his test flying was some years ahead of mine and, therefore, before the digital revolution which was becoming more and more pervasive during my years of flying.

Bill Waterton has written a book which encapsulates the way in which the UK aircraft industry had developed by the end of the Second World War. It had grown enormously as a result of the war with many separate firms with names which, alas, will only be remembered in the future by lovers of aviation history – Avro's, Hawker's, de Havilland, Supermarine, Handley Page, Vickers to name but a few. Waterton was drawn by chance into this scene when he was asked to join Gloster's and while painting his personal reminiscences he records some of the successes and failures of UK aviation in the post-war years.

His test flying career finished in 1955, just as mine was beginning at Avro's, though I well remember his famous name surfacing when I was attending the test pilots school a year earlier. In fact our paths unknowingly crossed just briefly as we both took part in the 1953 Coronation Display, he demonstrating the brand new Javelin fighter while I was tucked away leading a flight of Venoms.

Like many test pilots of his era, Waterton's flying hours were built up during the war, in his case as a flying instructor and then in Ferry Command. Unlike most of his fellow wartime pilots he realised that if he was to continue to fly in peacetime he needed 'specialized knowledge' and so, trying to take advantage of jet aircraft entering the R.A.F., he volunteered to join a new unit equipping with Gloster jet fighters. In fact he was turned down for this job but in 1944 was offered what turned out in the end to be a better alternative, Fighter Command's Air Fighting Development Unit based at Wittering. It was there that he was able to fly the very latest German and Allied piston-engined fighter aircraft and it was also there that he learnt that aircraft did not always fly or perform as advertised. He made the discovery, as I did later, that firms, the media and nations did not always tell the truth about their products. By 'dog fighting' these aircraft he learnt the true strengths and weaknesses of not only the Spitfires, Mustangs, Tempests and Mosquitos but also the captured Messerschmitts and Focke-Wulfs.

In 1945 the unit was transferred from the 10,000 ft grass airfield at Wittering to Tangmere with its 6,000 ft concrete runways, thus permitting the arrival of the long-awaited jet aircraft, the Gloster Meteor, together with some German jet aircraft for comparison. Clearly, Waterton must have been a very capable pilot and very effective in writing reports on the advantages and disadvantages of all these aircraft since he was selected to join the flight which was just forming to break the world's airspeed record using the Meteor; his piloting colleagues to be were the officer commanding of the flight and later air correspondent for the *Daily Telegraph*, 'Teddy' Donaldson, and a trainee test pilot, later chief test pilot of Hawker's, Neville Duke. After some hard work in 1946 Donaldson did manage to break the record with Waterton just two miles per hour slower in an inferior aircraft; they were both awarded the Air Force Cross while Duke had to wait a further seven years to break the record, this time in a Hawker Hunter.

Waterton had been trying for some years without success to get a permanent commission in the R.A.F. and so when Gloster's approached him in 1946 to join their firm as a test pilot at the princely salary of £1,000 a year he accepted their offer. From that point onwards he describes in his book the challenges and difficulties of being a firm's test pilot. He became chief test pilot a year later with his salary raised to £1,500 a year and, bizarrely, was then offered a permanent commission in the R.A.F. which he decided to refuse. In the next ten years he flew and supervised the development of the Meteor, from the Mk IV to the Mk VII trainer and the Mk VIII R.A.F. fighter, as well as the Javelin night fighter with its unusual delta planform and high tailplane.

As I can vouchsafe from personal experience, a test pilot has to do a lot more than just test aircraft. I recognised Waterton's descriptions of selling and delivering Meteors to Belgium, Holland, the Argentine and Egypt. He also had to go out to Argentina to try to sort out why there was so much difficulty re-assembling the aircraft. Again I felt his description of flying at Farnborough S.B.A.C. air displays in the '50s was very perceptive, highlighting the first class organisation, the often competitive nature of the individual displays and the excellent supervision of the flying displays by the flying control committee headed by the local group captain supported by non-participating S.B.A.C. test pilots. thus ensuring maximum safety both for the pilots and for the public.

Because Gloster's was a member of the Hawker Siddeley Group, Waterton had the opportunity to go to Canada to develop Avro Canada's brand new CF-100. The aircraft had quite a few problems, the most serious being weakness in the wing design and he had a couple of flights which might well have ended in disaster. He spent almost a year longer in Canada than planned helping to sort matters out and in the end the aircraft proved to be a first class all-weather fighter for the Royal Canadian Air Force but disappointingly was never bought by the United States Air Force despite a lot of effort.

The book is full of fascinating stories of test flights that did not go as planned and other flights that should have been routine but clearly were not. This, of course, is very understandable when one considers the lack of navigational aids at the time combined with aircraft that often had very little fuel. Furthermore, overseas flights were not coordinated as they are today with backup radar coverage. In addition Waterton, very interestingly, is prepared to admit when he made mistakes like landing without lowering the landing gear in a CF–100.

Equally interestingly, Waterton is not afraid to discuss financial matters, which included not only sales agents' fees but the financial rewards of being a test pilot in the UK; he makes it clear that the job, at that time, despite the high risk level particularly in the years he was flying, was incredibly poorly rewarded. There was no shortage of pilots wanting to be test pilots with the imagined 'glamour' associated with the task and firms did not really appreciate that all pilots were not equal, so that they fell into the trap of paying as little as possible and thus perhaps not attracting all the best pilots – though many pilots were prepared to work for very little reward for the sheer thrill and challenge of the job. It is interesting to compare the French approach to test pilots and test flying where the pilots were paid much more than airline pilots, unlike the U.K. at the time where the airline pilots were unionised and the test pilots were not.

Waterton was a typical example of the immediate post-war test pilots. Because of the war they had had no opportunity for advanced aeronautical training. They were invariably chosen because of their superior piloting skills which were very often needed to deal with unexpected and undesirable handling features of the aircraft they were flying. In later years the advent of computers enabled the designers to test their designs in flight simulators, so that the flying characteristics were more predictable and the need switched to test pilots who were trained engineers and computer literate.

Waterton effectively highlights well the difficult role of a firm's test pilot, very often the 'meat in the sandwich' between the certification and customer pilots on the one hand and the firm's chief designer on the other. The firm is always trying to get the aircraft certificated as quickly and as cheaply as possible with the advertised level of performance or capability, so that any problems that the firm's pilots discover are bad news for the firm and cost money to correct. Consequently, it was particularly easy in Waterton's time for a test pilot to become 'persona non grata' with the firm's management since the test pilot always tended to be a bearer of bad news if he required changes, almost an in-house Cassandra. There was always a tendency then, without the multitude of recorders now fitted to test aircraft, for the firm to prefer not to believe their own test pilot and take no action until the certification pilot or the customer criticised the product; adopting this approach, of course, proved more costly because, by then, more aircraft had been built and needed modification. The chief test pilot therefore needed to be an expert in handling the design office so that his criticisms were valued and not ignored. Clearly Waterton found this role difficult and the impression is gained that towards the end of his tenure there were problems in this area. However he made a very shrewd move in recruiting Dicky Martin before he retired because Dicky, who eventually joined Avro's and became a dear colleague of mine, not only did a spectacular job clearing the spinning characteristics of the Javelin for the R.A.F., a not inconsiderable task with associated high risk, but was also an expert in quiet diplomacy.

Waterton knew that to be a test pilot he had to have something special to distinguish himself from other pilots. However, having made the cut and got on the test flying ladder one got the feeling reading the book that ten years later his career somehow had ended rather abruptly and that all his very valuable work was not really appreciated at the time. Waterton's book gives an absolutely splendid insight into test flying in the post-war years discussing many of the issues that still exist to-day, though inevitably modified with the advent of the digital and computer revolution and the vast size of modern-day development teams. In the end however, one always has to remember that it is people who design aircraft, manage projects and eventually make things work and this is clearly reflected throughout in Waterton's very illuminating and intriguing book.

Tony Blackman
Marlow, 2012

PREFACE

I wanted to call this book "Suck It And See", for in those four words is the essence of experimental test flying. Despite the calculations and tests which go on before a 'plane takes the air for the first time, in the end it boils down to just that. But, as you see, I didn't get my way with the title; I failed in that small matter, and failure is the real tale of this whole book. For I feel that I have broken faith, with myself and with my fellows.

I experienced in the R.A.F. many aeroplanes which were far from what they should have been. I didn't like it, for a 'plane which is not as practical and foolproof as human endeavour can make it is a dangerous, or potentially dangerous, vehicle. It is not appreciated just how a switch in the wrong place, resulting in an awkward movement by the pilot, can mean, in case of emergency, the difference between "going in" and "getting away with it". I left the R.A.F. and went into the aircraft industry, wide-eyed, as a Sir Galahad eager to slay and put down the avoidable evils which beset my beloved mechanical birds.

For seven long years I did my best to put good, safe, world-beating aeroplanes into the hands of young, inexperienced squadron pilots of my old service. Despite a few minor triumphs I don't think I accomplished a lot. I didn't accomplish or bring into being a fraction of what I consider I ought to have achieved. Looking round the aeronautical field, I don't think others have done a great deal either. But I am out of it all now and so in a better position to say what I feel I must.

This book is not likely to make a mark as a treatise on how to win friends and influence people. If that were its object it would have been done on more stereotyped lines. As it is, it is just possible that a number of people will have quite a lot of things to say about what is written here. That is expected; I've been called a lot of things, very occasionally to my face—very frequently *à derrière*. It will be surprising if this book provokes anything new in invective.

Those who know aviation will recognize the evils to which I refer. The shirking of responsibility of jumped-up people in a jumped-up game. The old school tie habit of covering up and passing the buck. Greed, inefficiency, and "couldn't care less". All these are rife in larger quantities than aviation will stand. And the end product is that mess I have so often identified on the mortuary slab. But the blame is rarely laid at the right door. For aeroplanes do not fail as a result of "just one of those things". They fail or crash because of incompetence anywhere along the line, or through over-confidence, carelessness, laziness, un-truthfulness, or the pursuit of wrong ideals. They fail or crash because somewhere, sometime, someone has boobed, and for no other reason. All these are human faults, and to my knowledge no other of man's creations will show up human failings, wherever they be, more quickly or more ruthlessly. The sea is a hard master; the air even harder, for its foibles are so often invisible. All this being so, I often wonder if it does not account for the lengths human beings go to try and deceive one another over air matters, for aviation has bred a prize collection of art-ful dodgers. The measure of their deceptions is invariably directly pro-portional to their lack of first-hand knowledge of flying. And, "you cannot shoot a line about an aeroplane with impunity", for those who fly will not be fooled; "suck it and see" will reveal the truth.

WHAT DOES IT FEEL LIKE?

The E1/44, looking curiously frail and ugly, glittered in the pale, March sunlight: a tubby, trunkless, silver-winged elephant of an aeroplane, the vast air intakes of her jet engine serving as ears. She had no official name although Ace had been suggested. I called her the Gloster Gormless since she was so heavy for her single Nene engine.

Some six years and a reputed half-a-million pounds had been spent to create the Gormless from an aircraft designer's good idea. Now, according to the boffins, there was no reason why she should not be the jet-propelled answer to Fighter Command's prayer.

First, however, she had to be tested. As yet the Gormless had never left the ground.

I walked apprehensively towards her across Boscombe Down airfield. I had never before taken up a brand new prototype, and I knew that despite the cold logic and calm dispassion of scientific theory, things can—and do—go wrong between the blueprint perfection of an aircraft on the runway, and that same aircraft once she is airborne. Adding to my qualms was my knowledge of E1/44's history, which would have sapped the confidence of more experienced, less imaginative men.

The first E1/44 had been severely damaged (and subsequently written-off) the previous spring—1947—when the lorry bringing it to Boscombe from the factory at Gloucester crashed into a stone wall. Later, during taxiing trials in September, '47 on a second prototype, there was trouble with "shimmy"—nose-wheel wobble—when it felt as if the entire 'plane was trying to swivel round the front wheel of its undercarriage. Then, as I sped the Gormless along the runway at about 140 miles an hour as though for a take-off, the front wheel, with terrifying suddenness, took charge of the 'plane. Oscillating from side to side, it tried to drag the Gormless both ways at once. It was impossible to control. I had experienced many rough rides in my time, but this was sheer Hell. The aircraft rattled like a monstrous *maraca*, chattering my teeth and scrambling my brain. It stopped only after I had braked and the speed was down to twenty miles an hour.

I climbed from the cockpit numbed and dizzy, but in one piece. The Gormless, however, for all the science that had gone into her creation, lacked human resilience and had literally shaken her front end to pieces. Back she went to the workshops for the winter.

While in the shops, a further defect was discovered: a fault in the mechanism which locked-up the wheels in flight. Had she flown that previous September there was every possibility that having raised the wheels I should not have been able to lower them again, and would have been compelled to put down the Gormless on her belly. Not an encouraging thought that an engineering fault might have caused a prang on my first flight in a completely untried aeroplane.

I resumed taxiing tests at the end of February, '48, and still found the front end a bit "soft". The trials showed the Gormless to be lethargic and gutless. She accelerated slowly, and took a lot of stopping. The vital elevator control which governed her climb seemed heavy, and the rudder poor. All in all, she gave every promise of flying like a brick—especially when compared to the lively, virile Meteors, abounding with zestful energy, which I had previously piloted, and in which I had helped to break the world speed record. She had only a 1 to 3 thrust (weight ratio) against the Meteor's 1 to 2—1lb. of thrust (or push) for each 3 lb. of total weight.

At last the Gormless was as ready as she was ever likely to be. But I was hardly enamoured of her. As I walked towards her on that March morning I hoped my face hid my feelings. Frankly, my entrails did not.

A large crowd had gathered to see me take-off: Gloster V.I.P.s, Ministry of Supply officials, R.A.F. observers, scientists, technicians and ground crews. Once in the cockpit, things were not so bad: they never are when thoughts are replaced by actions. I checked instruments and switches, and felt more sure of myself. Then as the engine broke into a cockpit-muffled hum, rather like the sound one hears inside a power station, the last of my fears vanished.

Taxiing out, lining-up on the runway, and re-checking controls were old stuff. The radio call from the tower, however, was new: "Gloster One"—not Pink Pill Four, my normal call sign. "Are you receiving?"

"Loud and clear," I replied. "Permission to scramble?"

"Gloster One—clear to scramble. . . . Good luck."

"Thanks. I can always use that."

I took a deep breath and released the brakes. The restless tremble of the engine was replaced by a smooth forward motion. The air speed indicator rose slowly—oh, so very slowly!—to the 60 mark. Over that speed the engine started to bite into the speed-packed air, and she became alive. At 140 the Gormless left the ground. I climbed gently to 500 feet and pressed the button which retracted the three landing wheels. There were two thuds as they locked home.

Then, without warning, the 'plane swung violently to the left.

In perhaps a tenth of a second, fear squeezed my guts into a hard, iron ball and jellied my spine.

Then it was over.

Instinctively I had pressed hard on the rudder pedals in an effort to straighten the 'plane. At the same time I remembered the absence of a third thud as the wheels came up: one for each. I looked at the indicators and saw, to my renewed consternation, that the starboard one glowed red. The starboard wheel was locked neither up nor down, but might be anywhere in between.

For a terrifying moment I wondered what the devil I should do. In a proven 'plane a cockeyed wheel was bad enough, but in a prototype the asymmetric drag might cause anything to happen. If I put the other wheels down, then raised them again—hoping all three would lock up—it could work. On the other hand, something was possibly distorted and twisted, and to mess about with the undercarriage was liable to aggravate the damage. Yet to go in with one main wheel up could turn the Gormless into seven tons of madly cart-wheeling, blazing metal and fuel. Better to try to land with all the wheels down—if they would go.

A reaction was beginning to set in, for I found myself trembling as I pressed the buttons—although my brain was ice clear. The warning lights glowed red, then green, as the other wheels went down. But the damned starboard wheel indicator stayed red. Unless something had gone wrong with the light, the wheel was not locked down and was, therefore, unsafe to land on. I rocked and shook the 'plane in an attempt to free it, but nothing happened. E1/44, however, was now flying straight, indicating there was no undue pull to port. At my radioed request, Boscombe's tower checked the wheels through binoculars as I passed overhead. They said that all seemed to be down.

I could do no more. I licked my lips, braced myself, and started to bring her in.

My speed was steady, I lost height gradually, and my direction was accurate. Seconds later the runway disappeared beneath the 'plane's nose. The red light stared at me like a malevolent, satanic eye. The suspense was agonizing.

At 120 m.p.h. the wheels touched the runway. Gormless trembled—and the light turned green. The blasted wheel had been down all the time, and the slight jar had made contact. The fault was in the electrical circuit. I ran along the runway wondering why the devil I had not taken to selling vacuum cleaners for a living.

Many pilots have come to grief at this point. Intoxicated with sheer relief, they have pressed the wrong button, pulled up the landing wheels, and have sunk their aircraft into the runway, and themselves into ignominy. I checked dials and switches, taxied off the runway, and shut off the engine.

I had flown my first prototype. It had been both frightening and exhilarating—yet strangely unsatisfying.

1
THE R.A.F.

As the war rolled into its fifth year there could be little doubt of its outcome. And with the approaching reality of peace, planning for Civvy Street gradually ceased to be an amiably nostalgic, escapist game. Elaborately self-confident schemes and exuberantly optimistic dreams slowly faded, to be replaced by sober, worried uncertainty.

I was as concerned as the rest. I had learned, long since, of "the best laid plans of mice and men," and I realized that although post-war Britain would be hungry for labour, it did not mean that everyone was going to get the job he wanted. There would be professions in which the labour supply would far outstrip the demand.

This was particularly true of flying. There was going to be no shortage of pilots. Thousands of young men had been taught to fly. When peace came, enough of them would want to join civil airlines to create an employer's market. Similarly, the R.A.F. had plenty of young pilots from whom to choose its small, peacetime complement. If it had any soft spot for those of us older, pre-war pilots still living, and who were hoping to get permanent commissions, it was keeping the fact very much a secret.

We old-timers felt—rather bitterly—that we were getting a raw deal. There were not so many of us left—men who had joined pre-war on short service commissions and who had been fortunate enough to live through Dunkirk, the Battle of Britain and the fighting that followed. We had borne the brunt of the early years of the air war, and it rankled to see how we were overlooked for younger men who had not come along until the tail-end of the fight when we had plenty of aircraft and the pattern of tactics had been standardized from grim experience.

My qualifications hardly put me ahead of the field. Apart from service with fighter squadrons, I had been a flying instructor, and had flown the Atlantic with Ferry Command. Finally I had been posted to Fighter Command's Meteorological Flight—a twenty-nine-year-old Flight Lieutenant with some 1,700 flying hours behind me. The records showed me to be, in effect, just another fighter pilot, and the fact that good fortune had permitted me to survive was an ironic disadvantage: by peacetime flying standards I was going to be an old man.

But pessimism is as fatal as brash over-optimism. Logically, I reasoned, my best bet was to acquire some expert, specialized knowledge which the bulk of the other chaps would not have. I bided my time, went about my duties, kept my eyes and ears open, and my chance came in the spring of 1944 when rumours went around that a unit of Gloster jet fighters was to be formed. Jets were startlingly new, and not only did they offer the opportunity of getting in on the ground floor of something exciting and revolutionary, capable of stirring the imagination, but they also presented me with an opportunity of getting that "something extra".

I applied for a posting to the jet fighter squadron, only to be told that there was nothing doing. There was, however, an alternative. Fighter Command's Air Fighting Development Unit (A.F.D.U.) at Wittering, Lincolnshire, would soon be getting the jets for assessment and could do with an experienced fighter pilot. Although I would have preferred the jet squadron, this sounded the next best thing. I accepted the posting to Wittering.

The Unit was self-contained, and responsible only to Sir Roderick Hill, Chief of Fighter Command. A vital part of its work were "Comparative Trials", when we made practical comparisons between various types of aeroplanes. This was quite distinct from the theoretical work done by the Ministry of Supply's pilot-boffins at Farnborough and Boscombe Down. These pilots were interested in "flying machines", we with "fighting machines". The scientists' conclusions (which took an unconscionable time) were theoretical, worked out with pencil and paper. We, on the other hand, were practical operators: if we wanted to compare a Spitfire with a Messerschmitt we flew them and found out. Not only was this somewhat quicker than elaborate, treble-checked statistical collations, but it was also rather more realistic.

We tried out one 'plane against another, friend and foe alike, noting the points which make an aircraft good or bad, including such things as speeds (level and in the dive), rate of climb, manoeuvrability, steadiness of aiming and firing.

The combinations and permutations were endless. We flew Spitfires of one type against Spitfires of another, Spitfire against Mustang, Spitfire against Tempest, Mustang against Mosquito, Mustang against Messerschmitt 109, Spitfire against Focke-Wulf 190, Mustang against 190, Spitfire against 109—and so on. Each in turn was the attacked and the attacker. Then the two pilots switched aeroplanes. After that another pair tried it—and still others if we had large disagreements in results.

We soon discovered the truth about aeroplanes and equipment. We learned that certain firms did not always tell the truth about their products. Fed with their figures, the press and radio, in turn, misled the public. We found that nations did not always tell the truth, either, and we became bitterly aware that the best looking tables and figures issued by the boffins would not save a pilot's neck in a tight spot.

There came the point when I could tell the nationality of an aircraft blindfold. Germans by the individual smell of their finish, Americans by the size of their cockpits and the batteries of electrical switches,

British by the functional crudeness of cockpits. Flying characteristics, too, were national. The Americans were heavy, slow in the climb, fast in the dive, slow to gain speed, generally gutless, with light-feeling controls which, unfortunately, did not really shift the aircraft very rapidly.

The Germans varied considerably. Some, like the Messerschmitt (Me) 110, the Heinkel 111 and the Junkers 88, were dead easy. Others, such as the Focke-Wulf 190 and the Me 109, could be tricky. The narrow undercarriage and small canted wheels of the Me 109 caused a lot of trouble in take-off and landing, and the lack of warning prior to the aircraft's stall must have cost many 190 pilots their lives. The jets which came towards the end of the war—Arado 234 and Me 262—demanded considerable concentration from their pilots, especially on take-off and landing. All in all, though, the German 'planes were good performers and some, if handled properly, superior to their British counterparts.

Like the Germans, our 'planes varied a lot. Simple to fly, they were rarely vicious and gave plenty of warning before "acting up". Light and powerful they were easy to "chuck about", but tired their pilots by virtue of heavy controls and marginal stability which necessitated constant attention to flying. They lacked the solid feel of the American and German 'planes.

From the information we gathered, we sent reports to Fighter Command. We told of our likes and dislikes, suggested modifications and improvements, and gave positive information about fighter tactics. Thus, if a Spitfire was attacked from above or behind, it was pointless for the pilot to try to dive away—the Focke-Wulf 190 was initially faster in the dive, and—piloting being equal—the Spit stood no chance. If, however, the Spitfire went into a tight turn and stayed there, the enemy could not compete. In a turning dog fight, too, the Spit could always see off the German. Things like that, and "never try to outroll the 190", we were able to advise the squadrons.

Meanwhile, the boffins attached to us arrived at their own conclusions—not necessarily in agreement with our own—with the use of slide rules and logarithms.

Since a fighter exists only to use its armament, we held numerous trials to test the accuracy and effectiveness of guns, fighter-bombing and rocketry. During rocketry and sight-testing trials on the Tempest V, I was afforded a revealing glimpse of the mentality of certain Civil Service scientists.

Attached to us for the trials were boffins whom we used as "accountants" to calculate and give a rough guide to sight settings, and to collate results. Results in the field were disappointing to the Air Staff and our job was to devise techniques and equipment to improve these squadron results.

At the completion of the Tempest trials, which cost some £10,000 in rockets and fuel (apart from time, and wear and tear on the aircraft), one boffin blandly announced that they would have to be repeated. It would be tedious to detail the technicalities, but it boiled down to the fact that the scientist had made an error in one of his calculations. It was a tiny, inconsequential mistake which only affected his pieces of paper and not the hitting of the target, and I pointed out that an appendix to our report would put it right. There was the devil of a row before I won my way, but I was shaken to see how the man was prepared to spend a small fortune in public money to correct an error which did not matter a tinker's damn one way or the other.

The Unit had its share of accidents although—remarkably—none were fatal. A senior officer smashed one of our Germans and dented a Spitfire, while another had an impressive list of Spitfires to his credit. My own record with the Unit was to be three aircraft in about a thousand hours of flying over some two years.

Despite official propaganda, the R.A.F. did not get along very well with the American fighter wing who shared the station with us. The Lightning squadron stayed at their end of the mess, and we at ours. The coolness was not due to criticisms of one another as flyers, but to fundamental differences of temperament. We felt rather bewildered by the Americans, for instance, when, after the loss of an aircraft, tears flowed with beer and their end of the mess bore a maudlin, funereal air that lasted a week. They, in turn, could not understand British distaste for public exhibitions of grief, and were appalled by the R.A.F.'s: "Poor old Mike went for a burton this afternoon. Let's have a drink on him."

At the end of 1944 the Americans left and the Unit became the Air Fighting Development Squadron of Central Fighter Establishment: a centralization of many specialized units in Fighter Command. The new set-up was not a happy one for us. It was too big and too impersonal, top-heavy with executives, and we suffered from its numerous, tortured birth-pangs.

Curiously, different R.A.F. units attract—or develop—distinct types of flyers, and there were frequent clashes of personality between the "wilder" pilots of day fighters and the more sober pilots and observers of night fighters: rather "superior" types, we thought—"We have Radar .... All frightfully technical, you know."

My service at Wittering was noteworthy for a particularly memorable leave—albeit nearly a fatal one. In the winter following D-day I went to see how rocket firing squadrons were getting on and to look up a Canadian school chum serving with a reconnaissance squadron in Europe. (Unfortunately, I did not find him—although a subsequent search led me to his unmarked grave by the side of a Flanders road.)

I flew over in a Spitfire, landing at Antwerp twenty minutes before a V.2 hit the runway. There was still fighting around the docks, and the city was under a non-stop V.1 and V.2 bombardment. With a couple of Canadians and a Norwegian I did the night spots which, incredibly, were still functioning. The next day, following a senior officer whose navigation went adrift, I was led over the Rhine bridges near Arnhem, where we were "bounced" by two patrolling Tempests, but managed to get out of their way in the nick of time. Their language was not very polite. Later we were shelled by the Huns from the Reichwald Forest, and saw German jets in action as they streaked across Vokel aerodrome dropping anti-personnel bombs with remarkable and dis-turbing accuracy.

My return to England seemed doomed from the start. The engine failed on take-off, and I was lucky to get down unscathed on the run-way. Then there was undercarriage trouble. By the time the plumbers put things right and I had flown to Brussels, I was overdue from leave. To add to my troubles the weather became atrocious. The Met. man at Brussels assured me that conditions were better towards the coast, and were perfect at Manston, my destination. So I took off with less than a mile's visibility, cloud keeping me down to under a thousand feet. I dodged the chimneys around Lille and flew on at tree-top level. Heavy mist appeared below, and as I raced on its grey fingers groped up and around me. I turned back, but it was even worse for the weather had clamped down behind and fog completely covered the ground. Things looked better towards the south, so I set course down the French coast. Soon visibility picked up and the ground fog cleared, though the cloud base was still low and it was raining lightly. Well south of where I should have been, I set out for England on a new course and started across the Channel far from happy about the situation.

I had no dinghy and my radio had packed up. Then I ran into heavy rain and began to be really worried. My fuel was low, visibility had again deteriorated, and I would have to put down. But where? I could not be sure of reaching England, and it was virtual suicide to go down in the sea. I was in a bit of a spot—until a desperate association of ideas made me remember the airstrip back of Le Touquet from which Jerry had thrown everything at us in the far-off spring of 1940.

I turned on to a reciprocal course—and found it. But I need not have bothered. The concrete of the runways was pockmarked with a lunatic pattern of bomb craters. I was now so far off my normal route that I had no maps with which to find Amiens to the east. Equally, Le Havre was out. I had been in the air for one and three-quarter hours, it was growing dark, and my petrol gauge showed five gallons: insuffi-cient to take me to Paris. There was only one thing for it. I picked a field sloping up the north bank of the Seine, put down my wheels to show the locals I was friendly, then moved the selector lever to bring them up again—in accordance with forced landing regulations. But by that time the engine was turning too slowly to bring them up, and I landed with them still down.

The field was soft and soggy. At the end of the short landing the Spit rocked gently forward on her nose and broke the tips from her propeller. I got away with a bruised knee. I was some ten miles west of Rouen, and was welcomed by the friendliest of peasants. For the first time in my life I got tight—on half a tumbler of Calvados, followed by two of Pernod: a mixture which proved a remarkable aid to fluent French.

Eventually I organized the collection of my aircraft and came home. I was "in the cart", but fortunately someone had just smashed up an Anson by following the same Met. information, so my adventures with the Spitfire went comparitively unnoticed.

In the spring of 1945 our base was transferred to Tangmere in Sussex. It was there that we got the long-awaited jet-propelled Meteors.

My first impression was a pretty poor one. Taxiing along the ground they were miserable weaklings, and it often seemed touch and go whether they would take off or not. In the air, however, they were certainly equal to our best piston-engined aircraft. After the noise of orthodox 'planes, they seemed to slide through the air in an uncanny, effortless silence.

The German jets we flew had an even worse take-off than the Meteor, but in the air they were far more efficient than the British, going much faster—up to 100 m.p.h. or more—on very little power.

Our stay at Tangmere was short. Questions were raised in Parliament about the noise of our guns, bombs and rockets over Pagham Harbour, and we were banished to the wilds of Norfolk's West Raynham—a pre-war Blenheim station where, in 1940, I nearly shot down a Blenheim in mistake for a Junkers 88. I took over as Officer Commanding, Flying, and after five-and-a-half years as a Flight Lieutenant was promoted to Squadron Leader.

But the situation was more depressing than ever.

The vast wartime expansion of the R.A.F. seemed to have left it without any overall peacetime policy. No one knew where he stood. Those, like myself, who hoped for permanent commissions, were disheartened to see themselves overlooked in favour of younger, end-of-the-war recruits, and wartime flyers signing on for short period commissions.

Discipline was dreadfully lax. Many senior officers, who had been magnificent war leaders, were dismally lacking when it came to administering a peacetime air force. Not merely did they fail to have the training, but they set no examples. Once I was on the carpet for failing to discipline an airman, and the C.O. wanted to know why. I told him that I lacked the heart to charge an airman for drunkenness when relatively senior officers could be seen publicly in a similar state. The C.O. agreed that the situation was not an easy one.

In the late spring of 1946 the C.O. called for me again, this time to introduce me to a small, blond, fresh-faced Group Captain: Teddy Donaldson. I was to join the flight he was forming. Its purpose was dramatically exhilarating—to bump up the world air speed record out of reach of the Americans.

The previous year Group Captain "Willie" Wilson had pushed it to 606 m.p.h. We knew the Americans were ready to have a go with a Lockheed Shooting Star, and our purpose was to stymie them. Donaldson, an ex-Hendon aerobatic pilot, was to have two flyers: myself, because of my experience with Meteors (which would be used for the attempt), and a young man training as a test pilot at Farn -borough—Neville Duke.

2

THE AIR SPEED RECORD

I was deeply conscious that every pilot in the Service would have given his right arm to be in my place. I felt immeasurably proud—and very aware that with all eyes upon me I faced heavy responsibilities as well as possible glory. Adding to my keenness was the knowledge that my showing might powerfully influence my chance of a permanent commission. All in all, the challenge was fascinating.

A little frightening, too. For we would be pushing the Meteor further than had ever been attempted. We would be entering new—and quite unexplored—realms of high speed flight. But we felt that we would be contributing, in a modest way, to practical knowledge which would stand the R.A.F. in good stead when there was general use of the "hotted-up" aircraft we were to fly. (Ironically, at the time of writing—*some nine years later*—aircraft of comparative speeds are still not in large scale service with the R.A.F.)

The High Speed Flight was formed on June 14, 1946. During the preceding months there had been preliminary—theoretical—preparations. Results of the previous year's speed record were studied, and the three-kilometre course surveyed. It lay over the sea between Rustington and Angmering on the south coast. Our base was at Tangmere, and a great deal of organization was necessary, embracing such bodies as the Royal Aero Club and the National Physical Laboratory, and involving the setting-up of balloons, Air-Sea Rescue craft and the timing arrangements. Much time and money was to be spent on the attempt—and not without reason.

The jet aircraft was still something of a novelty. The Germans and Italians were out of the market; France, at that stage, was not in the running, and the Russians—as now—were a large question mark. The immediate future of jets rested with the Americans and ourselves. To break the world speed record with a jet (held pre-war by the Germans at 479 m.p.h. with a specially modified Messerschmitt) meant great national prestige and vital export orders.

Just after the war the R.A.F. took the record with Wilson's 606 m.p.h. in a Meteor. This was a formidable increase, but it was felt that it would not stand for long.

During the war we had sent Meteors to America—as we sent so many things we did not have the facilities to develop ourselves at the time. And, as usual, the Americans proved themselves brilliant developers of other people's ideas. The result was the Lockheed Shooting Star. With a fraction of the Meteor's power, it was almost as fast, and had far better control and endurance. We had wind that a hotted-up Shooting Star was ready to attempt to break Wilson's record, and we estimated that it could do it—reaching 615 m.p.h., which would give the 1 per cent. increase necessary for a new record.

In turn, British aerodynamacists and engineers reckoned that we, too, could raise Wilson's record and create a new one—but only just—out of reach of the Shooting Star. Their calculations and cautious confidence were based to a great extent upon a new, secret development of "Nimonic" material. Used in the manufacture of turbine blades, it allowed the engine to operate at a higher temperature and stress—and so develop more power without damaging itself. Moreover, this wonder material added no extra weight to the engine.

The aircraft for our 1946 attempt were virtually the same as those used the previous year—long-winged Meteor IV fighters. The engines. however, were to be more powerful than in the standard model, and it was calculated that if the temperature was around 70 degrees we could achieve a peak of 623 m.p.h. Warmer weather would suit the aircraft even better—although not the engines—but we dared not expect too much from an English summer. For all-round performance 70 degrees would be ideal. In any event, we were certain that if we raised the record to between 612 and 615 m.p.h., the Americans could not raise it the further 1 per cent.—about 7 m.p.h.—necessary for a record recognized by the F.A.I. (Fédération Internationalle Aeronautique).

Since the course and the aeroplanes were new to us, we were given three standard, but not yet in service, Meteor IV fighters with which to practice. In them we learned the feel of the 'plane and the best approach to the course. With the blessing of Donaldson, I carried out detailed trials similar to those I had done at the Central Fighter Establishment. They concerned acceleration and deceleration, radius of turn, climbs and dives, etc. In the report later submitted were a number of hitherto unknown technical facts about high speed flight which I hoped would be of some use to the R.A.F.'s future planning.

Between the formation of the High Speed Flight in June and the actual attempt there was much work to be done. The course was laid out with marker buoys, height indicating balloons were sited and flown, and complicated timing apparatus installed. It was arranged for Mosquitoes to fly at each end of the course to watch that we did not exceed the maximum permitted height of 300 metres—just under 1,100 feet. The timed runs had to be done at under 246 feet. (As it turned out, our actual runs—two in each direction to negate wind effects—were at between 80 and 120 feet.) An Air-Sea Rescue Walrus was laid on to pick us out of the sea if we crashed. (A kind thought, but at our speeds the sea would be like concrete, leaving precious little to pick up.)

Pounding up and down on our practice runs was a strain on both pilots and aircraft. We were flying our hacks at Mach numbers .78 to .79 (i.e. .78 and .79 of the speed of sound) and the buffeting and bouncing we took felt exactly like skidding down a flight of stairs on our backsides.

These practice flights were not without their troubles. One day on a westerly run at 580 m.p.h. and 50 feet, my Meteor swung with disquieting abruptness to the left. Instinctively I pushed my right foot on the rudder to straighten her, and automatically reduced speed. Something was obviously wrong with the port engine, yet a glance at the dials revealed temperature, fuel pressure and revs. to be as they should. What the devil was the matter? Back at full power she was almost right again, but when I reduced r.p.m. a nasty vibration set in. I shut off the port engine and landed on one.

The port engine was in a dreadful mess. Some of the Nimonic turbine blades were broken. Others were nicked and bent. These had been damaged by broken chunks of the impeller vanes which had torn through the engine, ripping out its bowels. At first we thought that a bird had got in the way—a source of constant terror, for at our speeds a seagull would hit the light alloy framework of the 'planes with the devastating effect of a cannon shell.

But there were no signs of feathers or blood to support this theory. I was baffled, although—as is usual in such cases—the ground personnel suspected that I had made a bloomer and was keeping back the facts. Then, while my port engine was being replaced, similar damage was found in Neville Duke's 'plane.

It did not take long to discover the cause of the trouble. Parts of the Meteor's wing skin were secured by what is called "blind riveting". This is employed when the riveter cannot see both ends of an inserted rivet. A special "blind" rivet is used. The tiny steel mandrel heads from the "blind" ends of these rivets break off during the riveting and on this occasion had worked their way into the engine compartment to be sucked into the compressor of the engine. Bits of the compressor broke off and went through the guts of the engine to smash the turbine blades. This could have caused complete disintegration of the engine and—possibly—the aeroplane. The problem was solved by the use of a sealing fabric which prevented the entry of foreign bodies into the engine compartment from the wings. It became a standard factory modification on all Meteors.

Our engines were noteworthy in another way: one nearly swallowed Group Captain Donaldson. He was poking around the front of a 'plane while the engines were running, when our engineer officer, Squadron Leader George Porter, saw him rapidly disappearing into the starboard air intake. With a shout Porter flung himself, rugger fashion, at Donaldson—or what was left of him. The engines were shut down and he was extracted safely, although very bruised and green. He was lucky to have had Porter around—others have been killed that way.

In August the two Meteors for our attempt arrived, although the special engines, which we held, had not yet been fitted. The 'planes were beautiful jobs. The gun ports had been removed, and immaculate paintwork gave them a glass-smooth finish. All cracks, dents and holes had been filled to prevent eddies of fast-moving air which would not only reduce speed, but could cause severe buffeting and bad handling. The air brakes were locked closed for the same reason, and the crevices where they fitted into the wings filled and painted. Without these brakes we would take time slowing down.

With the coming of the aircraft, Donaldson dropped a bombshell, and an embarrassing situation arose. He announced that only two pilots would fly in the record attempt—and he was one of them. That left Neville Duke and myself in a bit of a spot. We had taken for granted that all of us would have a go. There was no reason, I thought, why we should not (one of us handing over his 'plane to the third man) and our performances in the hacks were close enough for it to be said that we had equal ability. Relations became strained. Finally, for reasons of which I am ignorant to this day, I was selected as the second pilot. Neville's face did not conceal his disappointment, and I could understand how he felt.

Group Captain Donaldson had selected EE 549 for his own aircraft. Mine was EE 550, a lovely craft, easy to fly, docile, and as smooth as silk. When I got her to high speeds, however, I found that she had one serious fault. At 580 m.p.h. her port wing started to go down. It happened gently and slowly—but with relentless firmness. No matter what I did, the 'plane rolled to the left. Donaldson agreed that the port aileron, which governed rolling to port, was "slightly out", and she went back to Gloster's for rectifications. A couple of days later EE 550 was back with the assurance that she was now right on the top line, and that she had been flown in level flight over the Severn at 615 m.p.h. When I flew her again, however, I could see little change. But Donaldson ruled: "No more adjustments—they might spoil things, and time is short," so EE 550 and I had to do the best we could. I only wish I had known then as much about the vagaries of Meteor ailerons as I do now.

It was soon after this that George Porter's alert and practised eye spotted further serious trouble—basic defects in the very structure of the two 'planes. The fuselage under the wings of each aircraft was distorted—crumbling inwards as though sucked by a giant vacuum—and the framework was bent and buckled.

Feeling rather browned-off, and wondering what would happen next, I flew both aircraft back to Gloster's. Rarely have I known them to get down to a job with the alacrity and speed with which they rectified those two 1946 record-breakers. In a couple of days I flew them back to Tangmere, doing the trip in EE 549 at 620 m.p.h.—although the wind was doubtless helping. How I wished she was mine for the attempt: no strong man stuff was needed there.

At last everything was ready—bar the weather. The special engines were installed, and we were surprised that only one spare was available. Then occurred one of those stupid incidents which would be comic if they were not so infuriating. Royal Aero Club officials—whose presence was necessary for the recognition of a new record—decided to make a Blimpish issue of the fact that I was born in Canada. Could a Colonial compete in a British team? There was talk of my being withdrawn. With all the civility I could muster I pointed out that I regarded myself as British, a fact confirmed by my passport, and that if it had been good enough for a Canadian, Lord Beaverbrook, to get the U.K. its warplanes, then it was good enough for me to fly the damn things. The matter went no further.

August chilled into September, and the elements conspired against us. We knew the Americans were ready for their attempt, and to get in first was vital. It was almost certain that the country which made the second attempt could not raise the other (if successful) by the necessary one per cent.

We wanted smooth air and—above all—heat. Heat to drive apart the air's molecules and raise the speed of sound. Since we expected to achieve Mach .82 (.82 of the speed of sound) the higher the speed of sound the faster we could go. But instead of hot sun and warm, calm air, we had cold, fog, rain and drizzle, with low clouds on the hills, and mists which made a grey, bleak oneness of sea and sky. We grew bored and stale. Yet we dared not leave Tangmere in case a slight break came, and we could not undergo further practice since our special engines each had a limited life of six flying hours.

On top of this the whole show was beginning to wear thin because of the disproportionate publicity we had received. In an understandable effort to bolster Britain's post-war prestige, the Press had built the High Speed Flight into something approaching the super-collosal—hyperbole encouraged by official quarters which should have known better. My own feelings were to do it first and trumpet afterwards.

By September, however, the enthusiasm of the newspapers—determined by other events and, perhaps, a feeling that the attempt was not going to be made—was wearing thin. So unpromising was the weather, in fact, that Duke flew my hack Meteor to Prague to show the Czechs. He took Prague by storm, and the shy fighter ace came back enthusing over the Czechs who had dined and wined him and had bestowed upon him their Military Cross.

He returned to find that during his absence there had been a slight break in the weather, and that the attempt had been made.

Saturday, September 7, was cloudy and promised rain. The thermometer reached 59 degrees, far below the temperature needed to get the best from our aircraft. We were told to stand by at 4.30 p.m. "just in case". Sir James Robb, C.-in-C. of Fighter Command, to whom we were directly responsible, was on hand with his retinue.

A grudgingly tiny adjustment had been made to EE 550's aileron trim in an attempt to stop the troublesome pulling towards the left, and at 4.30 I took off to check it and to have a look at the weather. I ran over the course once each way at high speed. The aeroplane still put down her left wing, but the course was pretty smooth despite the weather. On landing I reported: "I should say this is it—but I'd like to put on a bit of aileron trim to pick up that port wing." Nothing was done about the wing, however, and at 5 o'clock Donaldson took off to check my weather report. He agreed that conditions were reasonably favourable, and the attempt was on.

Donaldson went first at 5.45. Seconds later we saw him streaking towards the sea from Pagham Harbour, south of Chichester, the howling engines leaving thin snakes of black smoke. He landed fourteen minutes after his take-off. He had used two hundred and eight gallons of fuel, reported a fairly smooth trip, and appeared well pleased.

As I aligned my 'plane down the runway, drizzle lowered the visibility. I checked engines and cockpit, opened the throttle, and left the ground at eleven minutes past six.

Once in the air my stomach stopped misbehaving, and there was no room in my mind for anything save the business of flying the Meteor.

I retracted the wheels quickly, before air pressure, caused by high speed, prevented them from closing properly. I checked the lights to see they were securely locked, for a leg going out at high speed could throw us out of control into the sea. By this time I was already over Chichester, three miles from the aerodrome. My height was within the regulation 1,100 feet, my speed 520 m.p.h. I put over the stick and went into a long, sweeping turn towards the sea on my left. Practice was paying off, for by keeping the speed the same as then I turned naturally into my line of landmarks. This was an enormous help since it was essential to line-up accurately for the run over the course. To deviate on the run would cut down speed—and it would be difficult to re-align the aircraft.

As I turned I saw a Mosquito patrolling out to sea, alert to observe my height. Parallel with the shore again I increased speed and passed over the end of Bognor Pier at 800 feet doing 570.

I opened the throttle fully to 15,200 revolutions per minute. The engines bit into the densely packed air, chewed and swallowed it, then spat it out at supersonic speed through the red-hot tails of the jet pipe nozzles.

The need for intense concentration eliminated imaginative thinking, but I felt tensely confident.

The air speed indicator crawled up to 580. *The cow's putting her port wing down—as usual,* I thought. *Behave yourself, you slut . . . .*

590. That blasted wing's getting worse . . . . Come up, you—My wrists and triceps ached in protest as I gripped hard at the stick with both hands. *Nothing for it now but to sweat and hope . . . .*

600. I'm at 150 feet now . . . . That's Littlehampton ahead . . . . They're expecting big things from me, down there. . . . I hope to God my grip holds . . . . The speed was causing the Meteor to buck and shake, and it was now taking all the strength of my two hands to hold her level. It would have been an uncomfortable enough ride without the need to fight a dropping wing.

Ahead of me a line of barrel-like buoys bobbed in the sea marking the course, as precision-straight as cat's-eyes on a midnight road. To my left were Flight Lieutenant John Eveson's height-marking balloons, Gertie and Ermintrude.

God Almighty . . . !

I was in trouble—desperate trouble. We hit a bump, and the port wing went down, taking control of the 'plane. I could not bring it up again. The line of buoys flashed away to sea as we swung crazily inland at something over 600 miles an hour. Eveson's balloons were no longer in the corner of my eye, but dead ahead. *For Christ's sake get them down, John! . . . But how can you? How can you?*

I stood on the starboard rudder pedal and pressed with all my fourteen stones. For a second I saw a nightmare picture of searing, uncontrollable disintegration, and then—miraculously—my unorthodox use of the rudder did the trick—pulled up the left wing. We skidded past the balloons and shot out to sea in a slight climb. Speed fell, and I was able to take one hand from the stick.

As I chopped back the throttles a muddle of thoughts rushed through my mind. *I've given them a thrill down there . . . . Myself, too. . . . But what now? I must have another go. . . . Haven't Gloster's said the 'plane is O.K. at 615 m.p.h.—and more? . . . . Hasn't Donaldson said it's O.K.? . . . Who'll believe me if I go back? I'll never be able to hold up my head again. . . .*

I roared eastwards at 560 miles an hour. I knew that I'd had a fearfully close shave, but I did not dare to think about it. That could come later. I had no radio with which to inform those on the ground of what had happened: the aerials had been removed to improve streamlining. *Just as well. They can't order me down.* I turned near Brighton. The weather had deteriorated. Drizzle reduced visibility to three miles. But I knew the coastline well and could line-up without the necessity of miles of forward vision. And as I did so I thought of a way to beat that damnable port wing.

I found that I could jam my left arm like a rod between the side of the cockpit and the stick. As long as my palm (against which the stick was resting), wrist and arm held out, the 'plane could not alter course to the left. To ease the strain I would pull at the stick with my right hand, and to further help out the left wing lowness I could put on some right rudder trim. When we reached very high speed the stick would have to crush my palm, wrist and arm in order to take charge of the aircraft.

I tore westwards with the throttles fully opened. A glance at the instruments showed everything to be in order. My speed crept up to 590 ... 595 ... 600. At 605 miles an hour the agony was indescribable. It seemed as though every bone from the tip of my elbow to the palm of my hand was in the grip of a giant, remorseless nutcracker: this in addition to the spine-jarring bounce of the bucking aircraft.

My course reckoning was exactly right, which was as well, for I could make no correction under the circumstances. The Meteor screamed through the air faster and faster, and the pain in my arm became too agonizing to be felt.

Then, with a sudden flash, the marker buoys disappeared beneath us, and it was over.

I reduced speed and relaxed my arm. There was little time for elation at having beaten the aileron gremlins, however: three more runs had to be made.

They went without severe incident, but I have never been so grateful for anything in my life as I was when I passed Eveson's balloons for the last time. I turned towards Tangmere. After travelling at 600-plus, circling the airport at 130 seemed hardly to be movement. The landing was an anticlimax. I taxied through the drizzle to the hangar and allowed the engines to cool slowly before shutting down. My five runs, including the first frightening dud, had used 262 gallons of fuel, and had taken twenty minutes. It seemed longer. Much longer.

Then came the chatter.

"What went wrong on the first run? ..."

"You scared the life out of the timing people. ..."

"We thought you'd 'bought it'. ..."

"John Eveson would never have forgiven you if you'd damaged his girls. . . . "

I was never blessed with a great deal of tact—too much Irish blood, perhaps. "Damn John's great fat slobs," I exploded. "And as for people's claims about how fast they've flown EE 550 without the wing going down—it's complete, utter, undiluted piffle." I made my routine report and went to the mess feeling unutterably tired and miserably depressed.

A bath and a shave failed to restore my spirits. Things had not gone well. I knew that my ruse to beat the port wing had not been really satisfactory. I had been compelled to use rudder bias, and that would have set up a drag which would lower the speed. I had dinner and caught a bus into Chichester, where I saw a film. I returned, more morose than ever, sat in the anteroom with a pile of magazines and papers, thankful that I was alone. My arm was stiff and sore, and I thought: *What a crazy way to try to break records.*

At about midnight, Wing Commander McGregor, the Ministry of Supply officer in charge of the course, came in. He was full of high spirits. "Why the gloom, Bill, when a new world record's been made?"

His words only half registered. I was too tired to care, and went to bed. Next morning I gathered that there had been a party somewhere. The Sunday papers were full of it: we had bumped up the record to 616 miles an hour, I had done 614.

We all tried to better our efforts later in the month, but without success. This time Duke got his chance and flew my EE 550. He was unable to get round the course at high speed in her, and even after big adjustments he encountered considerable difficulties. This seemed to confirm that there was something wrong with the aileron and not with the piloting.

It had been an unhappy story, especially when it is remembered how much the success of the High Speed Flight meant, not only to the R.A.F. and Glosters, but to Britain.

I had been unlucky enough to get the 'plane with the faulty left wing. When I complained about her she was sent back to the makers.

They returned her after a couple of days with every assurance that she was on the top line. I flew her again—and found no change. An added worry was the fact that subsequent delays ate up the engines which had only six hours of flying life in them. Then Donaldson, doubtless motivated by the understandable worry that too much tinkering might completely ball-up the aircraft, ruled: "No more adjustments." So I flew her as she was—and nearly broke my neck and prejudiced the High Speed Flight in the process. But this was typical of what was to come later—pilots being compromised by technicians and publicity men to make good their ballyhoo. To default is to be discredited; to go on is to take needless risks. For instead of success, there might well have been prestige-shattering, disastrous failure, which would have been worse than if no attempt had been made at all.

I drew up the official report, submitted it to the C.O., and returned to the Central Fighter Establishment. Donaldson. Duke and myself were each awarded the Air Force Cross, mine coming as a bar to the one I was given in 1942 for my work as a flying instructor. Without wanting to appear ungrateful, I could not help reflecting that three months and fifty hours' flying at Tangmere had brought a recognition that had not come after a thousand hours' flying over two-and-a-half years as a test pilot with Central Fighter Establishment.

Recently I read how a pilot claimed to have got 632 miles an hour from a Meteor IV—even though our special engines had not been fitted. I've often wondered how it was done—and timed.

3

FASTER THAN THE RECORD

My search for that "something extra" paid off. During my stay at Tangmere I was approached by Gloster's who were looking for a man with jet experience to join them as a test pilot. Their offer had much in its favour, apart from the £1,000 a year involved. I knew their aircraft (indeed, they and de Havilland's were pretty well the only British firms producing jets in quantity at the time), I had met—and liked—many of Gloster's people, and I was extremely fond of the Gloucestershire countryside in which their works were situated. I made a few enquiries among R.A.F. and Ministry of Supply acquaintances, and learned that Gloster's were reputed to be a good, steady, old-fashioned firm, a little on the slow side, who had had a lucky break when invited to build jets, having little of their own on their plate at the time.

My first love, however, was the R.A.F., despite its faults, and I renewed my attempts to obtain a permanent commission. But the best I could elicit was: "Nothing yet, old boy. Why not take on for a further four years? You'd be certain to get a permanent commission then."

That did it. "If the R.A.F. can't make up its mind after nearly nine years whether or not I'm going to be of use to them," I said, "then it never will. Sorry chaps, but I'm off."

And I was. I accepted Gloster's offer, and on October 21, 1946, presented myself at their headquarters. I was greeted by the sales manager whom I had met at Tangmere, and during lunch in the canteen I saw one or two other people I knew. Bill Downing, the production manager, sorted me out in the same helpful way in which he had looked after me when I brought the two damaged Meteors from Tangmere, introduced me to people, showed me where the lavatories were, and turned me over to my new boss, acting-chief test pilot Philip Stanbury.

Including myself, there were three test pilots besides Phil. Of the other two, one was an ex-Squadron Leader, the other a Naval type. The ex-R.A.F. man's job was supposed to be service liaison pilot. This involved visiting squadrons and bringing their troubles back to the firm for discussion and solution. The Fleet Air Arm chap had come to Gloster's on loan from Hawkers to help out when Gloster's were short of pilots. Liking the district, the firm—or possibly seeing greater opportunities—he elected to stay. He had been through the test pilots' school run by the Ministry of Supply, and although his general experience was less than the others', he had a good knowledge of the requirements and techniques of experimental flying.

Stanbury was a handsome, boyish ex-Flight Lieutenant who had been loaned to Gloster's during the war by the R.A.F. He had done a fine job of work for them, including much of the early development work on the first Meteors.

During these trials a nasty incident occurred which badly shook Stanbury. At a high altitude and speed his perspex cockpit canopy burst, and although he suffered no wound, the terrible sub-zero cold during the greater part of his descent put Phil on his back with pneumonia. This had taken a lot out of him, and severely restricted his activities. He suffered continually from colds and sinus trouble, and was able to do little flying.

Despite these sicknesses, which kept him grounded for so much of the time, Phil kept his title as acting-chief test pilot. The appointment of chief test pilot was still held by a man who was doing no flying when I joined the firm, and had done little in the previous twelve months. He was occupied with something on the executive side.

The oddness of this set-up was rather baffling for a new boy, especially for one who, after eight years in uniform, was very much a stranger in civvy street. But I told myself that it was not for me to enquire into the firm's *mystique,* and took things as they were, accepting, too, that as a new boy I must expect to be given the simple jobs—mainly the testing of repaired and overhauled crocks.

It seemed to me that an element of self-satisfaction dominated the whole of Gloster's activities. What had been a small run-down organization—it was said that the Hawker-Siddeley group paid a mere £120,000 when they took it over some time before the war—had expanded under the needs of rearmament into a vast works with great, Government-built hangars. During the war, aeroplanes were produced by the thousand, and after peace came Gloster's still employed about 4,000 people.

The firm had spread in a curious triangle. The headquarters of administration and production continued on the original site at Brockworth, and the aeroplanes were built, assembled and flown from there. But with the advent of jets, take-off and landing distances increased beyond the safe capabilities of Brockworth's small airfield, so the firm took over a wartime R.A.F. 'drome five miles south of Gloucester at Moreton Valence, and new jets were taken there for testing before being delivered to the R.A.F. The third point in the triangle increased distances even more, for the designing side of Gloster's was more than fifteen miles from Moreton Valence, about three miles east of Brockworth, at Bentham. Prototypes of new 'planes were manufactured and assembled by hand at Bentham, then broken down again and taken to Moreton Valence (or to the M.O.S. aerodrome at Boscombe Down) where they were reassembled for their first flights.

In a nutshell: the ideas and prototypes came from Bentham; practical flying experiments were made fifteen miles away at Moreton Valence; the approved 'planes were translated into quantity production at Brockworth—and the 'planes were then taken back twelve miles to Moreton Valence to be tested before delivery.

In the autumn of 1946, Gloster's were in an unhappy state. From the 474 m.p.h. Meteor III the Meteor IV had been evolved. Externally the two types were similar, the biggest visible difference being in the ailerons and the larger engine nacelles which housed eighty per cent. more powerful Rolls-Royce Derwent V engines. The frame of this new Meteor was toughened up: it needed to be to stand the 'plane's extra 111 miles an hour speed, and a rate of climb that is still good by 1955 standards. But it was not strengthened sufficiently. After being built, further tests, following accidents, revealed it to be only seven-tenths as strong as it ought to have been. Consequently there was serious trouble.

Earlier in the year Gloster's chief production pilot had been killed when a Meteor IV broke up in the air at high speed. Investigation revealed that the 'plane was not strong enough for violent manoeuvring at the speeds at which its engine forced it along. The weakness lay in the wings and centre section, with particular emphasis on the rear spar. A hundred of these Mark IV's had been built, but they could not be issued to the R.A.F. until they were strengthened or their engine power reduced. Since the R.A.F. wanted faster, not slower, 'planes, this latter course was impracticable.

The real answer was a major redesign of the wings, particularly the rear spar and centre section. Instead, strain on the wings was reduced in the simplest possible way: five feet were cut off the Mark IV's wingtips. This certainly worked, but it had the disadvantages of raising take-off and landing speeds, reducing climb and ceiling, and of affecting turning ability at altitudes. There was also ground-braking difficulty, and less control of the aircraft at low speeds.

The modified Meteor IV's had flown, but were not fully proved, and when I arrived at Gloster's there was still a great deal of work to be done before they could be distributed in bulk to the squadrons.

Then there was another, almost equally serious defect with the Mark IV: the lack of a pressurized pilot's cockpit. The engines of the earlier Meteor III would rarely get them above 35,000 feet—if that. But the infinitely more powerful engines of the Meteor IV drove it to colossal heights for its time. These high ceilings—up to 50,000 feet—were of great value in fighter tactics. Above 37,000 feet, however, the air is so thin that even with oxygen there is difficulty in breathing: there is not enough pressure to force the oxygen through the walls of the lungs. Flying at up to 43,000 feet on oxygen is possible for short periods, but above that height—where it often was necessary to fly for test purposes and, of course, tactical reasons—life expires without pressurized breathing. One answer is the pressurized cockpit.

In this way the pilot lives, as it were, at a height far below his real one. Thus, if his altitude is 40,000 feet, the altitude in the cockpit is only 24,000; at 48,000 feet it is 28,000 in the cockpit. These contrived cockpit altitudes allow the pilot to operate in some comfort and safety. Certain reconnaissance 'planes were equipped like this during the war—including the sub-stratosphere Spitfire VII which I had flown— but, for some extraordinary reason, the first Meteor IV's were not. Without this essential modification the R.A.F. would not accept the new Meteors, so in addition to the wing clipping, cabin pressurization was also outstanding.

Gloster's production line was at a halt, and the machines already built were of no use to the customers. So there was little for the test pilots to do except fly already delivered Mark IV's back from R.A.F. maintenance and storage units for modification, bring in Mark I's and III's for repair—and hang around Brockworth and Moreton Valence.

I did not hang around for long. A week after joining the firm I was off to Paris. The French were reviving their famous pre-war international aero exhibition, and Gloster's, capitalizing on their world record, were showing two Meteors—the record-breaking EE 549, now fitted with the ordinary Meteor IV's Derwent V engines, and a Mark IV fitted with the new clipped wings. I was given EE 549 to fly over, while our ex-R.A.F. man took the Meteor IV, and the former Naval pilot carried spares in a de Havilland Rapide.

The Channel crossing was quiet, as was the trip across France as far as Abbeville. Then, south of Amiens, the visibility deteriorated. The clouds, which completely blanketed the sky, lowered, and I was forced down to a ceiling that varied between 1,500 and 1,000 feet, with a murky view ahead that extended no further than two miles. I found Versailles and Villacoublay aerodrome, but my destination airfield, Toussis-le-Noble, eluded me. Visibility was now down to a mile, and ceiling to five hundred feet, with possible landmarks blurred by haze. I managed to make radio contact with the R.A.F. at Buc and asked for radio homing. No one had previously done much flying in very bad weather at high speed, and I discovered that by the time Buc swung its loop aerial on to me, determined and transmitted my bearing and course to steer, I was miles ahead of my original transmission. Buc and Toussis were side by side, only some four miles from me, but I kept missing them, going round in a large square, unable to see either, because of the poor visibility. Finally I realized what was happening, cut my speed from 500 miles an hour to 160, and was able to obtain and follow Buc's instructions. I landed at Toussis with no more than enough fuel.

I was surprised and disturbed to learn that I was alone at Toussis. The other Meteor had taken off well before me, and if it had not already landed somewhere it would have run out of fuel. A young, very keen R.A.F. Flying Officer/Engineer arrived from Buc to take me under his wing, and was equally perturbed about the missing Meteor. He put out a general alert to the French and to R.A.F. units in France, but it was too early to take full "overdue action". We had a silent meal, at which we were joined by the pilot of the Rapide, none of us really noticing what we ate. Deep gloom marked the afternoon—until a 'phone call came through from the Meteor's pilot.

His navigation had gone astray, and shortage of fuel had forced him to land in a ploughed field about thirty miles north-east of Paris and fifty miles from Toussis. He was unhurt and his aircraft apparently undamaged.

We set out in a lorry driven by the engineer officer. With us were two R.A.F. N.C.O.'s who were to keep guard over the Meteor. It was dusk by the time we walked through the mud of the field to find the pilot standing by a 'plane with a posse of gendarmes glad to hand over their charge to the N.C.O.'s. Our chap was both angry and sorry for himself, determined to have his 'plane refuelled next day and attempt to fly her off. But the state of the ground into which the wheels were well and truly sunk made it doubtful whether the aircraft would move, far less fly. Closer inspection showed that the axles had been damaged, and the rough landing might have strained the undercarriage. A thorough check followed by repairs would be necessary before the Meteor could be flown with safety.

We left her in the care of the N.C.O.'s and made for Paris. En route, we debated how to get the Meteor to the Palais for the exhibition. In the end it was decided that the best thing would be to remove her wings, give her new axles, and tow her from the field, over the roads, and through the streets of Paris. That, in fact, is what was done, the actual transport providing a headache for the engineers who had a hell of a time measuring the width of roads and the clearance of bridges.

The weather clamped down: bitter cold, alternating between dense fog and heavy mist. But there are worse places to be stranded than Paris, and we made merry for three days. Then the weather improved, and our job of delivering the Meteors over, we flew back to Gloucestershire in the Rapide. We nearly froze in the unheated 'plane, but revived ourselves in the club on the edge of Lympne airfield where we cleared Customs. Leaving Lympne it grew murky, the radio failed, the 'plane began to ice-up and our pilot did not know where we were. We put on our parachutes, but luckily we did not need them—we broke through the weather and found ourselves near Swindon.

The *Paris Exposition* opened in early December, and the three of us went over for it: this time by Golden Arrow. The show was a great success. Gloster's were very much in the limelight, and bemeddalled dignitaries from all over the world came to see the famous Meteors. As the man who had been in the record flight I was eternally press-ganged by Gloster's salesmen into telling potential customers the wonders of the Meteor.

We had only three days in Paris, but returned again just before Christmas when the show ended and the aeroplanes were ready to be brought home. We went in the Rapide, piloted by the bloke who had force-landed the Meteor. So help me, he nearly did it again. Over France we had our usual bad luck with the weather. Visibility was poor, and our pilot started going more north-east than he should have done. A long time passed without us seeing any familiar landmarks, but our man was obstinate. Finally we convinced him that the Seine must be south of where we were, running roughly east-west. Since it also ran through Paris, it followed that if we turned south until we found the Seine, then turned left, we would find Paris. He finally accepted our argument, and we were proved right.

The servicing people had been optimistic when they said the 'planes were ready to be brought home. It took a few more days to prepare the Meteor IV, and there was no hope of EE 549 being serviceable until the New Year. It was decided that Jimmy Bridge, the ex-Naval pilot, would take the Meteor back and I would return with the other pilot and some of the ground crew in the Rapide. On the way home, Jimmy made a new Paris-London record at an average 520 m.p.h. Unfortunately the weather thickened and he went a bit off course. But for this his speed would have been much higher. A great pity and a rotten bit of luck.

My shuttling across the Channel came to an end when I finally went to collect EE 549—and was asked to try for the Paris-London record. There was considerable muttering in certain quarters about the new boy doing this, but I had my orders from the general manager, and that was good enough for me.

Our French agent, a tough, ex-Maquis boy, arranged for Paris to see the EE 549 in the air before I went home. To my amazement and delight, the French authorities gave me *carte-blanche* to "beat-up" the centre of the city—something unheard of in Britain.

On January 15, 1947, I flew on a reconnaissance, familiarizing myself with the centre of Paris from the air. I was thrilled at the opportunity, for it is not every day that one has a chance to "beat-up" a capital city the size of Paris.

But the spectacle would not be without its sobering dangers. All air displays have their attendant risks, yet here I would be performing acrobatics around and among tall buildings, flagpoles and overhead wires. I proposed to fly along the Champs Elysees at a height of only fifty feet between the walls of the buildings. The slightest error there could mean the slaughter of spectators as the Meteor fell among them. Timing and precision were essential, for my approach to the Champs Elysees would not be across wide, open fields, but a downwards sweep from over the Arc de Triomphe. At over 600 miles an hour such flying did not allow for even an infinitesimal margin of error.

Later in the morning the 'plane was checked and refuelled for the display, scheduled for noon. The press and radio had really gone to town. and were full of *La Grande Manifestation,* and *Le Meteor, L'Avion Plus Vite Du Monde.* Paris, it appeared, was seized by that crazy fever for lunatic spectacle peculiar to the French, and the streets would consequently be packed with intrepid Parisians anxious to be scared out of their wits. It was very flattering, but a heavy responsibility.

After a final check, I climbed into the aircraft at Toussis at 11.50 and prepared to take-off. As I opened up the engines for take-off a terrific explosion shook the 'plane, and a great sheet of flame shot from the entry to the port nacelle.

My heart quite literally missed a beat. I frowned and swore—then recollected that I had heard of these explosions in Meteor nacelles. They were generally believed to be due to mechanics leaving off oil tank covers, or to the vapour from tiny oil or fuel leaks in the nacelle. This vapour, com bining with air, made an explosive mixture which could re sult in a giant backfire sufficiently violent to blow panels off the engine compartment.

Supposing, however, that there was a different cause? What if something was radically wrong? Perhaps there was damage that would not show itself until I was over the dense crowds in the Champs Elysees...

My predicament was a nasty one. Common sense told me that I should return to the hangar and cancel the show. Yet I was desperately anxious not to disappoint the French. The whole city was waiting for me to come screaming over the roof-tops, and I hated the thought of letting them down. Further: to call off the show would reflect unfavourably on both British aviation and Gloster prestige. Weighing heavily on the other side of the balance, however, was that fear of turning the Champs Elysees into a slaughterhouse. At my take-off position at the end of the runway where the bang had occurred I set the brakes, left engines running and got out for an inspection.

As I climbed from the cockpit I wished to heaven that the decision was someone else's. I looked carefully at the out side of the 'plane. As far as I could judge, the engine was undamaged. I turned the problem over in my mind—and decided to go through with the display. I confirmed that all was well in flight before starting the display.

I roared in from Versailles at high speed. From then on things happened very quickly. I dived over the Arc de Triomphe and down the Champs Elysees at tree-top level between the canyon walls of the buildings. At well over 600 m.p.h. the stately avenue is not very long, and it was necessary to pull upwards by the Place de la Concorde to avoid hitting that stone toothpick, L'Obelisk. Up ... up ... up at full throttle from 50 feet to more than two miles. I slowed in the upwards rocket to 150 m.p.h. as seconds later I was upside down on top of a loop.

I repeated the low runs in each direction, fast and slow, right way up and upside down, only 50 feet above the Champs Elysee pavements. Acrobatics filled the gaps between the runs—upward rolls, inverted climbs and aileron turns. The row must have been frightful as the Meteor scorched and shrieked over the city and between the buildings. For blurred split-seconds I glimpsed the halted traffic and up-turned faces. The noise would certainly be shaking them up. It was bad enough in the open.

Other parts of the centre of Paris came in for attention, among them one which nearly brought us down: the Eiffel Tower.

It was a strange and wonderful experience to fly round the great structure looking up at the people on the galleries who were waving to me. But with an icy stab of fear I suddenly saw the sloping, almost invisible steel cables which guy the great tower. In dropping my height to fly around the tower, I had missed one by no more than twelve feet. A bit closer and the steel rope would have sheared off a wing as easily as a hot knife slices butter.

I left the tower in a hurry, sweating freely. I flew upside down from west to east along the Champs Elysees, and made an inverted climb over La Concorde. I dived in at speed from the east, and disappeared from view in a series of climbing rolls into the western sky. The show was over.

The French were delighted, and I had enjoyed a unique and exhilarating experience. Had the performance been given over London, I would probably have been locked in the deepest dungeon of the Tower for life.

I made a close study of the rules governing my Paris-London record bid. I would be timed from the middle of Le Bourget to the middle of Croydon. I would have to pass over each at hangar level for identification purposes, and notice would have to be given for officials and timing arrangements to be ready at both ends. The thought of ploughing through a busy civil aerodrome at high speed was a worrying one, but however dicey it might be there could be no disruption of normal airline traffic—and that was that. I flew EE 549 from Buc to get a good look at the Le Bourget area, for I wanted to be certain of my landmarks in order to pass over Le Bourget dead on course for Croydon. Otherwise fuel and time would be wasted.

I spent that night at the R.A.F. mess—a poor substitute for the Pigalle, but very comfortable, and cheaper. Next morning—January 16—I was awakened at some ungodly hour (about eight o'clock) by a young officer and a batman. I am a bad riser at the best of times—and I was feeling far from my best. I looked at them bleakly as the officer cheerfully announced: "The weather's just right for a record attempt to London, sir."

I groaned and covered my head with the blankets. I was in no mood for any sort of flying, far less an attempt on a record. But there was nothing for it. I climbed from the bed and shuffled to the window. Dawn was not far past, and there was still misty haze and fog about.

I bathed and dressed, and staggered downstairs for a reluctant breakfast. I felt awful; I always do at such an hour. At ten o'clock the sun showed itself, and the timing people were told to stand by. Towards noon everything was ready for the 208-mile sprint. I paid my mess bill, stowed my gear in the ammunition bays, shook hands all round, and climbed into the cockpit.

As I left the ground, things fell into their usual pattern for such occasions: as though I had been given a powerful shot in the arm, tiredness and tenseness vanished, my brain was clear and my hand steady, and only one solitary butterfly flapped in my stomach—the worry of possible failure.

Aeroplane, engines, radio and instruments were behaving perfectly. EE 549 had always been a lady—unlike her sister, EE 550. I climbed to 5,000 feet and picked up the landmarks that would put me on a straight course over Le Bourget to Croydon, allowing a couple of degrees to port to counter the south-westerly beam wind which would otherwise blow me north of my target. The rest was automatic: checking radio, engine temperature, fuel and oil pressures—all the essentials which, if forgotten, can lead to disaster. I set the gyro compass on the course by which I would steer—and the record attempt was on.

We went flat out over the Le Bourget hangars, the Meteor kicking, bucking and bouncing with twitching wings as the hard, compressed air slapped at her with sledgehammer blows. The engines shrieked as we climbed up and away towards the Channel. We were covering a mile in less than six seconds. We buffeted up to 12,000 feet and crossed the French coast some eight minutes after leaving Paris. So far, so good.

The seventy-odd miles width of the Channel gleamed like dull brass. There were no landmarks: my only guide was the compass. I decreased height and peered intently through the perspex canopy which protected me from a ten miles a minute gale outside. The turbines purred like contented kittens, and the engines bit deep into the fuel. I shifted my weight as the hard dinghy pack pressed against my back side—reminding me that aircraft can, and do, come down to "ditch" in the sea. But even as I thought about it I saw I would not need my dinghy—in a flash we had passed over Eastbourne and Hastings. Seven minutes had gone since leaving France.

Visibility was good—a miracle of good fortune for January over the hardest of all countries in which to navigate. The fuel was draining away rapidly, and I used gravity in a shallow dive to help me down towards Croydon. I radioed that I would not be long, and pushed the throttles to squeeze the last ounce of power out of the engines. The passage became rough, with far more bumping than I'd known at Tangmere. I hoped that no other aircraft would get in my way, for at my speed the controls were like lead. But there was no time to worry: about five minutes after crossing the coast Croydon was below and behind me. The sprint was over. I slowed down and turned to port. It had taken about twenty minutes.

My worry now was fuel. The engines had behaved perfectly, but what boozers they were! Since leaving Toussis they had swallowed 240 gallons of fuel, and I had barely enough to take me to Farnborough where I was to land and clear Customs. It would be a bit of an anti-climax to prang through lack of juice. I tried to get Farnborough on the radio, but the rough ride and buffeting on the last lap of the trip had "killed" the sets, despite their anti-shock mountings. Then I saw Farnborough below me, and knew that I would be able to glide in safely even if the fuel gave out.

Gil Harris, the Chief Controller, was waiting with his lads to greet me. I climbed stiffly from the cockpit, my demob chalk-stripe suit, never a Saville Row creation at its best, rumpled, my hair a mess, and needing a shave. Someone took pictures, and I went along to the mess while EE 549 was refuelled and checked over. Before I left again the news came through—twenty minutes and eleven seconds, to average 618.4 m.p.h.: two miles faster than the world's absolute speed record, and certainly a new Paris-London record by a generous margin. At one point we must have touched a top speed of 625 m.p.h. The Press were due, but I dodged them and flew back to Moreton Valence. None the less I read several startling statements which I was supposed to have made to reporters who had not even arrived at Farnborough before my departure.

There was nothing eventful about my trip back to Moreton Valence. The ground crews were pleased to see me and I took a taxi to Brockworth. The girls in the office were bucked over the record, but my fellow pilots evinced no wild enthusiasm. Publicity and sales were naturally elated, and they sent me to see the general manager who shook my hand and said: "Jolly good show. Waterton." I repaired to my small bachelor flat at Cheltenham, fed myself, then went to the pictures.

It seemed as though I had made a good start to the New Year in my new job.

4

THE METEOR STORY

At the end of the war Britain's jet supremacy was undisputed. The lead—given by Gloster's Meteors—was formidable, and lasted until 1948 when the Americans produced the Sabre, and Russia flew the Mig-15. At home the Gloster lead survived until about 1951, when Hawker's came out with their Hunter, and Supermarine's the Swift—but Meteors would undoubtedly provide the R.A.F. fighter backbone until 1956/7.

Gloster's overwhelming pre-eminence had more than prestige value: it was an enormous asset to the nation's export drive. Gloster's fighters were the best performers in the world. We held the records to prove it. The huge sums of money spent today by the aircraft industry on publicity and propaganda were an unnecessary expense for Gloster's in the immediate post-war years: our Meteors were willingly and freely advertised by press, radio and newsreels throughout the world. The Meteor was a wonder 'plane. It had the fastest speed and climb, the best manoeuvrability, and the highest ceiling of any production aircraft in the world.

De Havilland's were a good, pressing second, but their Vampire was not in the same class. Supermarine's had not made a conspicuous success of the Attacker—a jet developed from their piston-engined Spiteful—Hawker's were still building piston-engined Furies. The American Shooting Star, although more efficient in many ways, lacked the Meteor's all-round performance and was considerably more costly. Despite its many faults, the Meteor was far in advance of anything the rest of the world could offer.

Immediately post-war, the smaller Allied countries and the wartime neutrals were in a bad way aeronautically, particularly in terms of military 'planes. Lacking aircraft factories of their own, they were dependent on obsolete fighters—some ten to twenty years old—and war-surplus piston-engined jobs. America dished out many of these latter crocks almost free of charge, but there was a catch with most of them: spare parts were not included in the *largesse* and had to be bought at very high prices. None the less, America's military and civil representatives, acting as political and commercial advance guards, made great rodomontade of their charity.

The job of the Gloster salesman was to persuade these nations that their defence rested, not upon clapped-out war-surplus, piston-engined 'planes, but with jets—Meteor jets. Our men travelled the world in the wake of the free publicity, together with representatives from Rolls-Royce (makers of the Meteor's Derwent engines), for our interests were indivisible. The customers came to us, as well, when our products were on show at Radlett (where the first post-war S.B.A.C. displays were held) and at Farnborough.

None the less, aircraft selling is not an easy job. A Government wants to make certain of what it is buying before signing a cheque running into millions of pounds. So Gloster's built a special 'plane to accompany the salesmen; a civil, gunless version of the clipped-wing Mark IV. It was painted in dazzling carmine and white and looked very effective.

The civil Meteor—registration number G-AIDC—was built early in 1947. Since her maiden trip was to be a flight to Scandinavia, an important job was to increase her two-tank fuel capacity of 325 gallons. A third tank holding 180 extra gallons was fitted under the centre fuselage, but even this was not enough since, at low altitudes, the Meteor drank a gallon of paraffin for each mile flown. So Gloster's added 200 gallons more, contained in two bomb-like tanks, one under each wing. After tests to see if these additional tanks affected control of the aircraft, the Air Registration Board issued a Certificate of Airworthiness for what I believe was the world's first-ever civil jet aircraft.

One day late in March, while these preparations were going on, I was called for by the general manager. As I walked to his office I tried to guess what was on his mind, and endeavoured to recall any bloomers I might have made. His news was an absolute bombshell. "Waterton," he said, "I should like you to take over as chief test pilot from the first of next month. You've got a free hand to make what changes you like, and your salary will be £1,500 a year."

It was not surprising that a new chief test pilot should be appointed, for Stanbury—by then confirmed in his post—was still a sick man, unable to devote to his job the energy it demanded. What was unexpected, however, was my being selected to succeed him: the other two test pilots had been at Gloster's longer than I, and were still inclined to regard me as the new boy. I could only hope that there would be no friction. I was excited to think that I had been given a free hand—but it was to take seven years before I fully realized the irony of the date of my appointment: the first of April.

By way of a consolation prize, one of the other pilots, the senior in terms of service with Gloster's, was given the plum of showing the Meteor in Denmark, Sweden, Norway, Holland and Belgium. The combined Gloster-Rolls-Royce party was supported by a Rapide loaded with mechanics, and a mobile workshop and spares department housed in a cream-painted war-surplus lorry emblazoned with red Meteors.

The Rapide returned at the end of March—alone. The Meteor had been left behind at Brussels—in a somewhat bent condition.

It transpired that after our chap had flown the aircraft to show its qualities, selected pilots of each country were invited to have a go themselves. Apart from one or two heart-tripping moments, it worked well—until Brussels. There an experienced Belgian pilot came to melancholy grief. At more than 550 m.p.h. one of the landing wheels fell partially extended from its recess. (All the Gloster aircraft I flew were prone to undercarriage up-locking or door trouble.) The aircraft was thrown out of control, and did a series of upward rolls to about 10,000 feet. Only the fact that the Meteor was pointing upwards when the wheel came down saved the 'plane from complete disaster.

The gloom of the Gloster party was deepened by the coincidental presence on the airfield of a de Havilland party who were waiting to demonstrate their Vampire. The Belgian brought the 'plane safely to earth, and as it touched down it seemed as though things were going to be all right. The red Meteor sped along the runway—accompanied by fervent de Havilland prayers that it should prang. It did.

There were delighted grins from our rivals as the offending leg collapsed and the Meteor swung to a crumpled halt in a cloud of dust. The Belgian was badly shaken—and lucky to be alive. For the Meteor's tailplane spar was broken, and the wings had bent under the strain of the violent upward rolls. The Gloster people were decidedly morose.

Then the Vampire took off, a Belgian at the controls, a smug de Havilland crew nodding encouragement and approval. But their expressions froze as one of the Vampire's undercarriage legs failed to retract as she climbed away.

The Vampire built up speed, and—to the horror of the de Havilland party—proceeded to perform aerobatics with one leg down. Our rival's consternation was matched by a noticeable improvement in the morale of the Gloster people, whose lips could be seen busy in wishful prayer. This time, as the aircraft landed, it was their turn to beg for a prang, for retractable undercarriages are not meant to be abused as the Vampire's had been. God, in his dispassion, answered Gloster's prayers, too. The Vampire buckled-up her leg, slewed round, and ended up alongside the wounded Meteor—almost in perfect formation. Amid

Gloster giggles a second badly shaken Belgian was helped to the ground by unsmiling de Havilland personnel.

It was a glorious day for "Made in Britain".

There were Customs difficulties when we made arrangements for bringing the Meteor home because, of all things, she was not in the same condition as when she had left the country, and they weren't sure it was the same aeroplane! Eventually she arrived in bits and pieces by road and sea: a sorry looking sight compared to the carmine beauty of a few weeks before. She lay, a sad wreck, outside a hangar until parts of her were taken to fly again in our Mark VII two-seater, and the first Meteor VIII.

While the tour was in progress, there was a distressing incident at Moreton Valence. I was spending the morning at Bentham with George Carter, our chief designer, discussing details of a new project, when we were interrupted by a 'phone call. It was the flying controller at Moreton Valence. He had joltingly bad news. A Meteor had crashed into some trees when coming in to land, and Jimmy Bridge, our ex-Naval pilot, had been killed.

I gave instructions that the wreckage should be guarded, the aircraft's documents impounded, and the news suppressed until we knew more. I rushed down to see the general manager. He had heard the news and was white-faced. I told him what I had ordered to be done, and sug - gested that we should go immediately to Bridge's widow—before the tragedy was revealed to her by the B.B.C. or some overkeen newspaperman who might hear of it. Tremendous mental anguish can be caused—and is caused—by clumsy, unthinking behaviour of this kind.

The breaking of such news is always a rotten job, and was to be my depressing task twice during my years at Gloster's. On this occasion I took with me a Gloster staff nurse and the wife of Rodney Dryland who had recently joined us as a pilot. (Mrs. Dryland was to be the recipient of the same tidings herself two years later. I was in South America at the time and, frankly, grateful for it on that score.)

We found Mrs. Bridge shopping in Cheltenham. She was awfully brave about it, as the women of men who fly usually are. Looking back over fifteen years, a cruel quirk of fate always seemed to select pregnant wives to have such news broken to them.

So far as we could tell, an engine had stopped while Bridge's aircraft was approaching the airfield. (Bits of grit, it turned out, were getting under a ball valve in part of the fuel system, and after further incidents, none, fortunately, as serious, modifications were made to prevent this.) Bridge had apparently glided in slowly, found he was going to under-shoot the runway, and opened up. But only the port engine responded properly, swinging the aircraft starboard towards a clump of trees. Had he chosen to throttle back and land short of the runway, he would prob-ably have escaped—although the Meteor would have been badly dam-aged. Instead, in the manner of many pilots, he had tried to save his 'plane. Lack of speed and obstructions prevented it from coming off. The fighter stalled, fell into the trees from about fifty feet, and was smashed to bits.

I was told that as chief test pilot it was my duty to identify the body. It was an unpleasant task, and the sight stayed with me. But the job had to go on, and more aircraft were waiting to be flown that afternoon.

Bridge had been a good friend of mine. His death was a personal loss.

As a result of G-AIDC's unfortunate end in Belgium, Gloster's de-cided that never again would they permit the pilots of customers to fly their aircraft solo. Instead, we built the Meteor VII, a two-seater dual-control version—the first jet trainer ever constructed. At the time there was no official requirement for jet trainers, and the firm had to spend their own money on an aeroplane. (The enterprise paid off, for some 750 Mark VII's were ultimately sold. Further, the Meteor VII sired a series of Meteor night fighters built by Armstrong Whitworth.)

I was in on the design of the Meteor VII from her earliest days, and I had constant rows about her.

I was appalled to discover, for instance, that this aeroplane, with a ceiling of nearly 50,000 feet, was to be put into the hands of young pilots without the pressurized cabin essential for life above 43,000 feet.

I protested and raged—but no one took any notice. It never has been pressurized, and in general service has to be flown below 40,000 feet.

Lack of armament was a further point of disagreement. Without either guns, or provisions for bombs and rockets—which fighters only exist to use—the Mark VII could only pretend to give her pilots gunnery and armament training. I felt that potential customers, seeing this, would turn to other 'planes in which to train their pilots. I argued, too, that customers with limited budgets would prefer to purchase trainers easily adaptable to an operational rôle.

My quarrels against the Meteor VII extended to details. A pet hate was the swing-over canopy of the cockpit. It weighed 264 pounds, and its locking mechanism proved unreliable. When the aircraft went into service with the R.A.F. the large number of accidents caused by these canopies substantiated my views.

The rear cockpit—the instructor's—was, to me, an incomplete affair, and the biggest bone of contention of all. It lacked such elementary instruments as fuel gauges, jet pipe temperature indicators, and—most vital—relight buttons with which to relight the engines in the air if their fires went out.

I remember vividly, even now, the stink I raised with the design department over these points—and how I was cold-shouldered out of the way. I recalled, bitterly, talk of "a free-hand", and when I later learned that the R.A.F. were interested in the machine I tried to prod the Ministry of Supply into doing something before it was too late.

My conduct was to be regarded as unprofessional, unethical, underhand—and more besides, but I felt I had a duty to my fellow pilots in the R.A.F. and wrote to the Ministry of Supply through their representative at Gloster's, the Resident Technical Officer (R.T.O.). Racking my brains, and as a desperate last resort, I turned myself into a "Barrack Room Lawyer", and found a technicality on which I could embarrass them and pin them down.

I reminded them of the R.A.F. order making the captain responsible for his aircraft. The instructor, I said, would be the captain of a Meteor

VII, and since he would occupy the rear—instructor's—seat, he would be in an unassailable position if there was an accident. Unable to relight a blown-out engine in flight, and with no indication of his fuel state or engine temperature, it seemed unlikely, I submitted, that a court martial would be in a strong position to convict any instructor who damaged his engines, failed to relight them, or crashed through lack of fuel—assuming the pilot survived to stand trial. My letter caused a furore at the Ministry, but it was still years before the rear cockpit equipment was supplemented.

In March, '48, I flew the Meteor VII from Brockworth to Moreton Valence for trials, taking with me another pilot to check the instruments in the second cockpit. We were up for forty-five minutes. The trainer flew beautifully and was full of verve. Without guns, ammunition and other fighter equipment, she was nearly a thousand pounds lighter, and this showed to advantage in many ways. Take-off and landing distances and speeds were reduced, climb and high-altitude manoeuvrability were better, and the improved streamlining of the canopy suggested that the Mark VII was a bit faster than the Mark IV fighter.

At Moreton Valence I was given an acid reception by the chief technician who maintained that I should not have had a second pilot with me for fear of putting out the 'plane's centre of gravity. I told him that I could hardly take him seriously since the aircraft was meant to be a two-seater, and that the extra person moved the centre of gravity only an infinitesimal distance forward, which was in the safe direction.

I was particularly glad that the Meteor VII gave no trouble, for I had just flown the Gormless—which was destined never to go into production—and my hands were full enough with two experimental 'planes to cope with.

I put in fifteen hours of flying with the Meteor VII in order to compile the report on her performance and characteristics needed when we applied for her C. of A.—Certificate of Airworthiness. Obtaining the Certificate turned out to be a Goon Show of a procedure.

Virtually all R.A.F., Naval and military aircraft fly without interference from the civil authorities—and manage to do so quite well. But when one of these service 'planes is taken and flown as a civil aeroplane—even without passengers—it is a matter for wonder that red tape doesn't secure it to the ground until it corrodes. For a government agency, the Air Registration Board (A.R.B.) enters the picture. They are a sort of aeronautical Lloyds, and for a fee decide whether to issue one of their C.'s of A., without which no civil 'plane may leave the ground. A year earlier, when G-AIDC was built, an envelope of design data and test results satisfied this esteemed body of pioneer airmen and shipping experts. But that would not do for the Mark VII. Instead they insisted that one of their own pilots should take her up for testing.

The officials, inspector and pilot arrived one morning at the end of April. I showed the pilot over the aeroplane and cockpit, went through our test reports and the pilot's notes I had compiled, and had a very pleasant lunch with him at Gloucester. I then went about my other duties, leaving the A.R.B. pilot with the Meteor, feeling confident that the 'plane would not let us down.

At the end of the afternoon I was puzzled to learn that the 'plane had not been taken up. The A.R.B. party were in the cubby-hole which served as my office, and as I entered I dropped my parachute and asked: "Has the damned thing gone unserviceable on you ?"

"No, no," said the pilot hastily. "'I'm just waiting for you."

"For me? Why? What's the trouble?"

"No trouble," he assured me. "I'm just waiting for you to come with me."

"Me come with you? Whatever for?"

Then—and only then—did it come out. "Well . . . You see . . . I've never flown in one of these jet things before!"

Such was the Air Registration Board in 1948!

The dual flight went off very well—with me in the instructor's cockpit doing the take-off and landing—and in due course our C. of A. was granted. But I was slow off the mark, for I neglected to demand an instructor's fee from the Air Registration Board.

The world's first jet trainer was painted in carmine and cream, with the registration letters G-AKPK. She became quite famous in her way. She flew to many countries, and carried more V.l.P.'s—including Royalty—than any other aircraft of her day. She introduced many to jet flying, including Mr. T. O. M. Sopwith (Sir Thomas), Chairman of the giant Hawker-Siddeley Group, who was in his sixties, and his son, Tom junior, then sixteen. Ironically, the Sopwiths seemed to be the only members of the Group, or its subsidiary firms, willing to fly in their jet creations. The Sopwiths flew in the trainer when it had done less than twenty hours in the air: after thousands of hours behind her, the Meteor VII still did not seem to inspire the confidence of other Hawker-Siddeley or Gloster dignitaries, for none would ever come for a flight.

After scores of demonstrations at home and abroad, G-AKPK was "called up". Her silver uniform bore the insignia of the Royal Netherlands Air Force to whom we sold her, and who promptly wrecked her in a training accident. I had flown her almost exclusively before the transfer, she had taken me safely over thousands of miles of foreign travel, and had left me with many pleasant memories. I was sorry to hear of her end.

Gloster's next venture of consequence was the Meteor VIII, a further modification of the Mark IV. I flew her in October, 1948, but it appeared that the Meteor's days as an interceptor fighter were numbered. Hawker's swept-wing Hunter "prototype", the 1052, and Supermarine's Swift—both superior in performance and speed to the Meteor—were now flying, and had been promised to the R.A.F. in quantity at an early date. (Actually the R.A.F. did not receive its first delivery until 1954.) The Javelin was little more than an elegant design study, and it seemed as though Gloster's were going to face a barren period with nothing to build. Then a scheme was produced both to keep the workshops busy and to extend the life of the Meteor.

The idea was to employ the Meteor in a tactical rôle as a ground-attack 'plane. Most fighters end up in this capacity as they become *passé,* but I confess to a sentimental pang when I heard it was to happen to the Meteor. I recalled her days of defiant glory: the matchless world-beater of 1946. It was sad to think her moment of grandeur was passing.

Details of the ground-attack Meteor were copied from the American F-84 Thunderjet which had been relegated to similar duty. The most important modification was the addition of wingtip fuel tanks. These were necessary in order to give the Meteor the low-level range with which to reach the forward ground troops she would support, and they needed to be at the wingtips to leave the undersides of the wings free to carry bombs or rockets.

For some years I had badgered our design office to introduce these wingtip tanks instead of the underwing type used on the G-AIDC. After initial troubles, the Americans had been flying wingtip tanks for some years, but Gloster's designers did not appear to have faith in their own ability to make them work. Yet they were of technical, as well as tactical, value. They gave additional span, helped to straighten out flow at the wingtips where the air was disturbed, and, when full of fuel, relieved upward bending of the wings in flight.

Wingtip tanks presented the designers with two problems: making certain they would come off cleanly when jettisoned, without hitting other parts of the 'plane, and ensuring that they were right aerodynamically. This latter point had been the cause of many wing failures in America, for unless the shape of the tank and the angle at which it was attached were correct, "lift" from the tank would produce torque in the wing structure and the wings would twist from the aircraft.

But if the Meteor was to be a ground-attack aeroplane with any useful range, the tanks had to be fitted. Put to it, our boffins finally solved the problem as the Americans had done. The new 'plane could carry 1,000 gallons of paraffin, and weighed about 22,000 pounds. It still got off the ground in 1,200 yards, and would stay up, at height, for four hours. She flew well, even when loaded with twenty-four 100 pound rockets.

The new Meteor was called G-7-1. Less eye-catching than her predecessors—she was painted in service silver instead of carmine—

she had a chequered career. She was flown brilliantly at Farnborough by one of my pilots, but interest in her oscillated violently. When the Hunter and Swift failed to arrive on schedule, it meant that the R.A.F. would have to put up with Meteors for some time to come. That, in turn, meant continued production of the orthodox Meteors, and the Air Force showed little interest in our ground-attack version. So we used G-7-1 for other work, particularly the sorting out of problems connected with the new spring-tab ailerons.

After that G-7-1 hung around the hangar until early in '53 when we heard that the Indian Air Force were interested in Meteors. The orthodox Marks lacked the range the Indians required, so we wheeled out G-7-1. The customer arrived, but left without buying. I learned that a decisive factor was the bad service rendered by another British firm who had supplied aircraft to the Indians. This put them off our aircraft industry, and although they came to have a look at the G-7-1, they spent their money on seventy of the costlier French Ouragon fighters.

Our 'plane did not do much of anything after that. She was eventually made into the blue machine seen at Farnborough in 1954. A two-seater front was added to the centre fuselage, but she retained her wingtip tanks and Mark VIII style rear and tail. A far as I can discover, she has been quietly forgotten except for use as a "hack".

For communication and ferrying work we ran our dear old Rapide. Bought at the end of the war, she was the last one made. We flew many hundreds of trouble-free hours with her, and she accompanied our tours to the Middle East, Scandinavia and Europe. She was supported by a Proctor V—also the last of her type to come from the production line. The Proctor was all right in the air, but left the ground in a rheumatic stagger and was awkward to land.

Having flown American light aircraft—including the Cessna, Navion and Bonanza—I am always depressed when I reflect how badly Britain fares by comparison. It is amazing and infuriating to consider how this country has fallen behind in the manufacture of good light aircraft.

Britain, pre-war, produced the Moth series, Vega Gulls, Bluebirds, Avians and Tomtits. Today, however, she is far behind the excellent light 'planes made by the Swedes, Italians, French, Czechs and Americans. The reason why we have never competed in the post-war field is a mystery.

Yet the Dominions, and many foreign countries, cry out hungrily for small, versatile, sports, personal, business and agricultural aircraft.

I do not know whether our manufacturers lack interest, vision, courage, or faith in their ability to produce a competitive small aircraft—or whether it is simply that their enterprise has been emasculated by safe Government contracts—but the fact remains that no British aircraft company will touch a design for a small 'plane without a firm order.

Red tape does not help, either. Although Rapides have flown passengers with incomparable safety for twenty years, these tough old bi-planes do not begin to meet the current requirements of the Air Registration Board. A Rapide undercarriage leg was produced pre-war, I was told, for less than £30. Today, a manufacturer assured me, the same article would cost £600 in order to contain the modifications demanded by the A.R.B. (It is true, of course, that some of the increase would be absorbed by the higher costs of labour and materials, but they would not bump up the price by two thousand per cent.) This is a typical example of the growth of red tape in the last twenty years. Yet I do not think the present crop of light and low performance aeroplanes are safer and less trouble-prone than those of pre-war years.

Engines for small 'planes are another problem. The Gypsy and Cirrus have been going for more than twenty years, and have been built in their thousands. They must have amortized their tooling time and time again—yet their cost is prohibitive.

I did my best to persuade Gloster's to produce a small 'plane. I made a lot of headway, too, before someone at a high level killed the idea. I gathered that Sidney Camm, Hawker's chief designer, had a light, all metal 'plane on his drawing board, and one light aircraft in the group was enough, apparently. At the time of writing, however, nothing practical has come of that, either. But, of course, the workshops weren't so busy in the early days of the idea.

I flew many other 'planes while at Gloster's, and one of them, which did not come into my hands until late in 1953, was by way of being a charmingly romantic and nostalgic interlude. She was a pre-war Gloster Gladiator—the same indomitable Gladiator as the famed Faith, Hope and Charity, the early defenders of Malta.

Ours was the last surviving Gladiator. She shared a Brockworth hangar with our Rapide: a forlorn, forgotten old biplane, dilapidated and dirty, cocooned with cobwebs.

When I became chief test pilot I did all I could to interest the management in the possibility of getting the Gladiator serviceable again. I tried to appeal to their business sense as well as their sentimentality by pointing out that if she was used as a "hack", doing liaison work with the Navy and R.A.F., she would both pay her way and be a good advertisement. But the general manager said that it would cost £1,000 to put her into shape, and that the firm was not prepared to spend the money.

I had flown the Gladiator pre-war and had formed a strong attachment to it. Every time I went to see her—which was often—she seemed to reproach me.

She was sold, finally, to an ex-R.N. pilot named Bellamy who ran a small air charter service from Southampton. By dint of scrounging, engineering ability and a free new engine from Bristol's, he got her into the air again. He displayed her at air meets, obtaining expenses and a small fee each time she was exhibited. But she was too costly for Bellamy to run by himself, and although I don't know for how much they sold the Gladiator in the first place, Gloster's bought her back for £500. Bellamy said they "stole her".

After her return, I had lunch with Bellamy, and learned that the Gladiator was far from standard. Without guns or ammunition she was very light, and she had no starter or electrical equipment. There was plenty of work that needed to be done to her, but the management would not spend a penny on the old girl. She was held together, literally, with string and wire, and the civil air authorities permitted her to fly only near the aerodrome. On request this was extended to inter-aerodrome flights for displays.

I was told that she must be kept as a museum piece, and flown only on special occasions when the authority of the general manager would be required. This permission was not granted until the autumn of '53 when I was told that I could fly her to nearby R.A.F. airfields for Battle of Britain celebrations.

I approached the Gladiator with a certain trepidation, for I remembered clearly that these old 'planes needed a lot of flying. As I climbed aboard and levered myself into the cockpit, thirteen years slid away. Once again I was in an aircraft with a personality. I experienced a feeling that lovers of old cars will appreciate. Here was something fundamental and real: an aircraft with a propeller, constructed of fabric and metal, and with an engine I could see working in front of my nose.

I checked what few instruments there were, and primed the engine as a mechanic wound a handle thrust into the cowling. The prop turned with slow jerks as he wound. After more priming with petrol, and winding—as though we were cranking a car—the engine fired, coughed, belched and stuttered into life. The starting handle was withdrawn and the prop swung lazily some four feet ahead of my face. The gentle slipstream brought a familiar, oily smell.

This was how I had learned to fly nearly fifteen years before. The seat shook, and the entire 'plane trembled as I opened the engine to check the magnetos. A gale, now, from the airscrew. It mangled my moustache and ruffled the ground crew who were draped over the tail to keep it down. *God,* I thought, *these old ones are rough on the backside.*

We taxied out, the tough old Gladiator wobbling on its undercarriage. Then, as we went into the wind, she was airborne—as light as thistledown. For a moment, before the technique of handling the old girl came back to me, the Gladiator took charge. I had forgotten how light she would be after the powered bricks of today, and we whizzed all over the place at a rattling 160 m.p.h.

I tried my hand at aerobatics—rolls and loops and dives. This was really *flying:* a joyous thrill I had forgotten during my past four-and-a-half thousand hours in the air. Rolls, twizzles, stall turns—it was incredibly stimulating. During a slow roll the engine stuttered and stopped. I thought: *God, a forced landing,* and rolled right way up. The engine coughed, spewed flame, and caught on again. I remembered—Gladiators did such things in the normal course of events. There was no need to worry.

There were low clouds in places, and it was showery. But it did not matter—not in the Gladiator. With the cockpit cover open clouds and rain were not enemies, but sweet manifestations of nature that made you feel clean and young and exuberantly detached.

The wind shrieked through the wires as we went down at Moreton Valence. We touched the runway doing less than sixty, and after a run of no more than two hundred yards came to a gentle halt. The entire population of the aerodrome seemed to rush out and surround us. Among them were two young R.A.F. pilots who had come to collect Meteors, and who stood with open mouths. Poor chaps: they had been born a few years too late to appreciate the joys of real flying.

I grew to know the Gladiator well. She developed many ailments in her old and tired bones. One day I brought her down with a hole burned in her front cowling. She had exhausted her hot gases against a weary and worn spot in the collector ring, and it had given up the struggle. But the defect was made good, and I went on to exhibit the Gladiator at Aston Down and Little Rissington.

I think, quite sincerely, that I was sorrier to leave that old aeroplane than anything else at Gloster's. She was a reminder of the days when aeroplanes had personal characteristics reflecting the loving care and enthusiasm of those who gave them birth, flew them and administered to them. The Gladiator was a tribute to her designer, H. P. Folland. It was said that Gloster's never built a 'plane with decent controls after he left them. I certainly never flew one which, for her time and age, was her equal.

5
TURKISH TOUR

Not long after I had taken over as chief test pilot I received a letter which, in retrospect, seems incredibly funny: the R.A.F. had finally made up its mind and was offering me a permanent commission.

I was not amused at the time, however. I was extremely disturbed and worried.

For eight years a permanent commission had been my one ambition. Because it had not been granted I had left the R.A.F. for another career. What was I to do now? Go back and take up where I had left off—or settle down in my new job? The decision was not an easy one, for there were attractions either way.

If I went back I would have quite a bit of seniority by the time my flying days were over, and would be found another job. In the ordinary scheme of things one could expect to retire in one's early fifties, a Group Captain, with a pension of £875 a year.

Before considering the R.A.F. further, I went to London for a medical, for I would need to be fit before they took me back. I passed A.i. I also wangled a quick look at my documents, and it appeared that I was a most reliable and excellent officer—although I did spot somewhere the phrase "a rebellious character".

Heavy on the other side of the R.A.F./Gloster scales was the fact that Gloster's £1,500 a year was far more than the Service offered at the time. I talked things over with the firm, and they pointed out that while they had no pension scheme, one was under consideration, and that in the meantime I was insured for £3,500—precious little consolation to a bachelor, however. Gloster's seemed flatteringly anxious to keep me, and I was assured that irrespective of pension schemes, insurances or anything else, I would be "looked after".

The way they put it, the proposition sounded good. I let the R.A.F. go by—and was to curse myself for doing so.

The summer of 1947 was a hectic one, for my work entailed more than the testing of aeroplanes. All flying activities came under my ægis, and I found much that wanted doing. Flying control, for instance, had to be reorganized and equipped; crash and fire procedures evolved and made workable. It was impossible to tell why these most urgent matters had not been dealt with by my predecessors, for they left no records or instructions. (It astounded me further to discover that millions of pounds worth of aircraft had left the works before my appointment without apparently any real flight testing records being made of their testing, performance and behaviour.)

I could not operate that way. So I worked late into every night drawing up orders and instructions to cover the various matters for which I was responsible. During the day these instructions were taken to the people concerned for discussion, and it soon became apparent that my written orders were not going to be popular. The reason was equally clear: if there are no written orders outlining duties and responsibilities, there can be no pinning down. Instead, any investigation arising from a mishap is lost in a tortuous, interminable labyrinth of buck-passing. No one takes the can back; everyone remains complacently inefficient.

I was obstructed, sniped at and hindered, but I would not—I dared not—tolerate sloppy, indecisive and undisciplined behaviour. Other businesses might be able to "muddle through", but in flying you pay dearly for carelessness—usually with a pilot's neck.

It was a tough fight, but I won it. For a time, anyway.

Meanwhile, business flourished. We were given an order by the Argentine for a hundred Mark IV's, and despite there being no two-seater jet training 'planes (the Mark VII had not yet been built) we taught a dozen of their pilots to fly them. I engaged two more pilots: one to replace Jimmy Bridge on the experimental side, the other to test aircraft as they came from the production line. A third pilot went to the Argentine to re-test the Meteors when they were reassembled after their sea voyage.

By this time a few Mark IV's with long wings were with the R.A.F. for an "early look-see", and we heard a number of disturbing—yet not surprising—reports of "incidents". Not surprising, because jet flying was still in its infancy, and service knowledge of its ramifications was meagre. It might sound melodramatic to say that each high speed, high altitude flight in 1947 was a journey into the unknown, but it is not far from the truth, yet, frequently our reported findings were not passed on to the squadrons.

These incidents resulted in a conference at Boscombe Down. In addition to Boscombe's test pilots and boffins, staff officers from the Air Ministry were in attendance, together with Ministry of Supply personnel. The artillery was lined-up heavily against Gloster's, who relied in no small part upon the defence and counter-attacks I could put up.

The Boscombe barrage was short, but formidable. The Meteor's stability was poor at altitude. It did not fulfil requirements high up. At 40,000 feet, even in level flight, it ran into compressibility (control difficulties when approaching the speed of sound) which ruined it as a gun and sighting platform. Its speed range was small between the stall (the speed at which the aircraft would not fly) and compressibility. Without concentration the pilot could be in trouble either way. Further, the Meteor would not turn at altitude anything as tightly as was desirable.

I did not refute their arguments. I knew well the Meteor's short-comings. Agreed, it was not all it could be, but the criterion was simply this: was the Meteor safe and useable? With what was Boscombe comparing the Mark IV? What else had they flown at 48,000 feet? Were they not comparing the Meteor with an ideal? Under equal conditions, was it not far superior to any other fighter? Did it not boil down to the fact that the Meteor was so far ahead of anything else that people did not know how to approach it?

It was quite a session before the day was conceded to be ours.

September, 1947, saw the second post-war show put on by the Society of British Aircraft Constructors. Held for "the trade" at the Handley Page aerodrome at Radlett, it was closed to the public—except for members of the Royal Aeronautical Society who held their annual garden party at the aerodrome on the final day.

I have good cause to remember Radlett, 1947. I nearly "bought it" there.

The Meteor and de Havilland's Vampire were the stars of the show, for most of the other aircraft were piston-powered. A prototype Attacker jet was on view, and seemed very fast in a low, straight line, but since it was not chucked about much it was impossible to judge its true qualities. The tragic, tailless de Havilland 108—prototype of the Comet—was also flown, and seemed full of promise, but the Vampire and the Meteor supplied the spectacle, fighting it out side by side with every trick in the aeronautical book. It was an exciting competition, with our publicity boys constantly urging me to "see off" our rivals.

The dicey incident occurred during that simplest of manœuvres: a loop. I had done it scores of times, and its mechanics were almost automatic. This time, however, I must have been carried away by the publicity department's infectiously enthusiastic: "Give us something to talk about."

I knew I could loop the Meteor in under 3,000 feet of air. On that day cloud completely covered the sky at 3,500 feet. That ceiling meant going over the top of the loop at no more than 150 miles an hour, or even as slow as 130—lest I built up speed so fast in the dive that I could not pull out before hitting the ground.

Reefing-up from a screeching dive, I shot upwards and over the top of the loop, not at 150 miles an hour, but at 200. Added to that I did not pull back on the stick hard enough initially, and was doing around 300 while still partially inverted—and accelerating rapidly as I screamed nose-down towards the ground only a couple of thousand feet below!

I pulled on the stick with both hands—and almost blacked-out. I could see only dimly through the greyness what was happening, but I could *feel* the seven tons of Meteor heaving and shuddering under me, partially stalled.

Then I could see clearly again: the railway line. . . . A helicopter. . . . Rows of parked cars . . . The vast crowd with heads instinctively ducked. As they rushed towards me I thought: *This is it.*

Numb with fear, I held on, pushed out a bit of flap to get more lift—and waited for it. But the Meteor responded. Her downward "mush" was arrested, and she levelled out. I climbed again, badly frightened, yet aware that the show must go on. *Well, they wanted something to talk about and, by God, they got it . . . . And a lot more than they bargained for . . . . So did I . . . .*

I learned later that I skimmed the ground at about 10 feet, shaking the parked helicopter, and half-hidden by the dust cloud I raised. I've never felt quite so confident in a loop since then—even at 10,000 feet.

Ironically, just before I went up, the Chief Inspector of Accidents had prodded the Meteor and said: 'Wouldn't do aerobatics in that thing, old boy. You've got some rivets pulling." My misjudgment had put the aircraft through the book, but there were no signs to justify the Inspector's pessimistic warning.

Although the Meteor was unquestionably the supreme aircraft of its day, and virtually unrivalled, Gloster's salesmen had a tough job. For they found themselves up against American "dollar diplomacy"—a new, post-war imperialism operating under the guise of "military and economic aid". In part return for this aid—which was motivated by nothing more altruistic than a desire to extend commercial and economic spheres of influence—the recipients were cajoled or coerced into accepting American weapons for their defence: even though better ones were obtainable elsewhere. This was a great stumbling-block to the British aircraft industry, as I saw for myself when I went to Turkey.

It did not need an Old Moore to recognize that the Turks were nervous of the way the Russians eyed the Bosphorus, and as we knew the Turks' Spitfires and Mosquitoes were ripe for replacement, it seemed reasonable to suppose they were in the market for Meteors: a belief encouraged by the fact that Turkey had substantial gold reserves (a reputed £50,000,000) with which to pay for our 'planes. (Gloster's reasoning may have been cynical, but fear is one of the armament makers' best salesmen.)

Consequently, in the spring of 1948, we set off to show the Turks our carmine-and-cream Mark VII—the two-seater built to replace the pranged G-AIDC. No jet aircraft had ventured so far afield, and the trip involved impressive preparations—thousands of gallons of paraffin fuel and specially fine filters, for instance, were taken in advance to five airfields along the routes.

Our Rapide, in its faithful rôle of workshop and spares department, went ahead of the Meteor, flown by Group Captain Leonard Stokes, of Rolls-Royce.

I flew to Tangmere with veteran mechanic, Freddy Cook, in the back seat, and on May 21 set out across the Channel for Marseilles. The weather grew dirty over France, the air brittle with thunderstorms, and heavy, solid cloud roofing the sky at 30,000 feet. We climbed to 37,000, but there were still thunderheads above us. "Static" from the thunderstorms made it impossible for me to use my radio transmitter—and then the most vital instruments packed up. Not one, but horizon, bank and turn, gyro compass, altimeter and air-speed indicator. It was against all the laws of averages, the thing which couldn't happen with separate sources of activation, but it happened. Fortunately, there was no real trouble, and we dropped towards Marignan aerodrome an hour and a quarter, and 620 miles, after leaving Tangmere. We had averaged just on 498 m.p.h.—aided by a bit of wind—which was not bad going.

The Rapide was already there, and after arranging for B.E.A.'s engineers to repair the Meteor (it transpired that dirt and broken pipes had caused the troubles) we drove to Aix-en-Provence, where the night was enlivened by one of our crowd shaving at dinner with a portable razor and later by his efforts to bathe in a public fountain. From there we went on to Turkey, via Italy and Greece.

I had been told to land at Gazi Izmir, a large, modern military aerodrome with good, long runways. As I went in I saw plenty of upturned faces, but no signal to land. Puzzled, I flew two miles further to Cumovasi, the civil aerodrome: a scruffy little place, populated by camels and goats, and with an indifferent runway. The Rapide was there, but its crew seemed to be pointing back towards Gazi Izmir. Completely baffled, I returned to the military airfield and put down. The Meteor was immediately surrounded by a crowd of gesticulating, babbling Turks. From the size and nature of the rumpus it was clear that I was on the wrong aerodrome.

When I proceeded to taxi away, however, I was immediately waved down. I had no idea what the devil they were playing at—but at this point an old, pre-war Mercedes dashed across the tarmac. In it was a Rolls-Royce man, resident with the Turkish Air Force, who told me that although I could not stay, I would not be permitted to take-off again for another half-hour.

Impatient at all this messing about, I learned that there was a road connecting the two aerodromes, and with the Rolls-Royce man as guide, I made the trip to Cumovasi overland. It must have been a startling sight: the carmine-and-white Meteor tearing along a Turkish country road in a great cloud of dust. Fortunately there was no traffic to bar the way, and my only fear was of kicking up stones which would enter the nacelles and wreck the engine. But we arrived safely, soaked with sweat in a forenoon temperature of 90 degrees.

Leaving the mechanics to refuel the Meteor—this involved pumping 600 gallons of paraffin by hand from 45-gallon drums—we were met by our Turkish agent's local man. He had arranged for us to stay overnight at Izmir and to fly to Ankara the following day, so we piled into a couple of cars and set off for the town.

The following morning I flew to Ankara. Central Turkey was very badly mapped, there were few landmarks between towns, and the towns themselves were queerly camouflaged by nature, blending almost invisibly into the brown earth on which they lay. I flew by dead reckoning, and was relieved to sight Ankara only one minute later than I should have done.

We were met by a Rolls-Royce representative who installed us in a hotel. His news was disturbing. Instead of enthusing over our arrival as we had expected, the authorities were exhibiting a "hard to get" nonchalence. We went to the Embassy to see H.M. Air Attaché, Air Commodore Wigglesworth.

It appeared—although Wigglesworth did not put it so bluntly—that the Turks were currently being wooed by the Americans, and that United States *largesse* was effectively shouldering-out British interests. To win Turkish friendship (and to increase their military and economic grip on the country) the Americans were practically making a present to the Turks of war-surplus, piston-engined aircraft, and since the Turks knew virtually nothing of the advantages of jets, they were not particularly impressed by our arrival. On the contrary, they were falling over backwards in their efforts to embrace the kind and generous Americans.

We reasoned that the only way of beating the highly organized, government-sponsored pressure tactics of the Americans was to show the Turks by practical demonstration that the Meteor was the world's best fighter—even though we were not in a position to supply them free of charge. We put on a little pressure ourselves, visited the Air Ministry and Defence Department, and arranged that on the following day we would put the Meteor through her paces before members of the General Staff. It was agreed that I would also make a flight over the city.

Ankara aerodrome was rough, and there was constant danger of blowing a tyre, or of kicking up stones into the engine. Even when airborne my troubles were not over, for there were tall wireless masts just south of the field, and at certain angles they disappeared into a background of distant hills. The greatest worry, however, was the buzzards which wheeled indifferently through the air and whipped past like small shells. But all went well. The Meteor rose to the occasion, behaved perfectly, and the buzzards kept out of the way.

The town received its share of thrills as I screamed over its rooftops. I was told later that the cave dwellers in the old part of Ankara insisted that Allah had returned from the skies. Among the thousands of upturned faces were those of the Russians on the roof of their Embassy: a great white building which was easily recognizable. I thought: *If you want to see this British 'plane close-up, you shall*—and passed over their heads so low that I could almost see them clench their teeth as they ducked their heads.

The display worked wonders. The Turks had never seen anything like the Meteor before, and were powerfully impressed. The radio and press took up the cry, and must have used every adjective of praise in the Turkish Thesaurus to enthuse over the Meteor.

I was asked to display the Meteor again to a further gathering of military top brass. Government ministers, and members of the Diplomatic Corps—excluding the Russians—were invited, too. At the end of the show seven Turkish officers, including two generals, flew in the Meteor, and as an ironic end-piece I took up Lieutenant Colonel Jim Ferguson, a member of the American mission. He had flown Shooting

Stars, but was not prepared for the Meteor's performance. I recorded in my log afterwards that he had been "shaken rigid". It was good to have been active at a time when the Americans were forced to admit the superiority of a British fighter.

The British colony was delighted with our success, saying that we had done a tremendous amount towards raising British prestige. But the Meteor's triumphal tour was not over, for our services were now much in demand. Military and civil authorities throughout Turkey were anxious to see "the red firebird", and it was agreed that we should display her at Eskiehir—their Cranwell—Istanbul and Ismir.

I arrived at Eskiehir on June 1, and performed a few aerobatics over the town before landing—always an effective method of rousing local interest. From here the Meteor, with Turkish officers in the second seat, went over several outlying towns. One in particular I remember, cradled at the foot of sabre-toothed rocks like the mountains of the moon. Whilst at Eskiehir we also took in a Mohammedan wedding.

The display went off smoothly, and we moved on to Istanbul. There we were involved in a "diplomatic incident".

The Meteor was displayed to President Inonu, whose son came up for a ride. I went on to make six different flights, but the jet still had not been seen by all the V.I.P.'s who were interested in her. So another display was laid on at Yesilkoy, Istanbul's civil airport. The authorities could not do enough for us, grounding both their own 'planes and a Swiss airliner to leave the air clear for me. Radio messages warned all other airborne aircraft to keep away.

Low clouds at 1,500 feet ruled out loops and vertical manoeuvres, and I was restricted to demonstrating the 'plane's agility and speed in the horizontal rather than the vertical. While flying inverted at a low altitude I saw, to my horror, that another aircraft was in the air—and dangerously close. I reefed round to avoid a collision and recognized the 'plane as a United States B-17 bomber. It was not in radio contact, and despite signals was determined to circuit the aerodrome.

Since the Americans seldom pass up a smart sales trick, I thought that I would take a leaf from their book. I tore round the airport and passed the B-17 three times before it had completed even half a circuit. It gave the watching Turks a practical demonstration of what the Meteor could do against a piston-engined bomber.

The bomber landed, and I completed my display. On touching down I was met by an excited crowd. A member of my ground crew, grinning broadly, said: "You haven't half upset the apple cart. There was an important Yank passenger aboard, and he's just exploded to the Press. You must have scared him."

Another also told me that the Americans present didn't like it. Understandably, for a lot of their time and prestige went down the drain as the value of their old crocks suddenly dropped considerably in Turkish eyes.

As we walked across the tarmac I saw from the corner of my eye two American officers approaching. When you see American officers walking erectly you know that they are on their dignity. I reduced this considerably by increasing my pace. To catch up they had to break into a trot.

A voice called: "Hey, fella—you flying that goddamned thing?"

We sauntered on, the advantage ours, for authoritative argument is difficult when you are running after someone.

The two Americans were naturally Colonels, and I gathered from their breathless, jogging outbursts that "their General" had been aboard the 'plane, determined to land irrespective of the Turks' injunction. General Hoag had been given a fright. . . . If their guns had been loaded they would have shot me down. . . . If it had happened in America I would have been grounded for life. . . . If——

I expressed regret at frightening the General but suggested that he was the one to apologize. After all, it was our show that had been interrupted, not his. I added that I was sure nothing had occurred that a drink would not settle, but the invitation was turned down.

We had an excellent lunch with H.M. Ambassador, Sir David Kelly: the epitome of charm and courtesy. After hearing our story, with diplomatic tact he left it to the Turks to sort out things with the Americans, although his wife said with a smile that she would appreciate it if we did not make Sir David's job more onerous.

The General's voice did not go unheard. He made *Time* magazine.

Our final display at Izmir's military aerodrome was highlighted by the General commanding telling us that he thought the Meteor far superior to any U.S. aircraft he had ever seen—despite persistent American attempts to denigrate our 'planes.

The trip to Turkey was not without its social life.

During our five days at Ankara we ate regularly at one of the world's best restaurants, a magnificent place run by a bald eagle of a man named Baba Karpich. In his seventies, he was a White Russian, and was reputed to have been a wealthy oil man from Baku until the revolution forced him to flee to Greece. He opened a restaurant in Constantinople, and when the town was taken by the Turks, his establishment came to the notice of Kemal Ataturk. The dictator was impressed and said: "Karpich, you will go to Ankara and open a restaurant. Except for State establishments you will hold a monopoly of the eating trade."

Ataturk is dead, but Karpich's survives. During the war he fed interned R.A.F. officers with the best in the house—despite their meagre cash allowances. "Other ranks" were not charged a penny. I was told that he had been recommended for an O.B.E., but that they had never got around to giving it to him.

Stokes, the Rapide pilot, knew of the old man's weakness for a good cigar—something virtually unobtainable in Turkey at the time. So we stocked up with £25 worth of Havanas *en route,* and presented them to Karpich as a gift. It was a master-stroke. At every meal he presented us with a large bowl of caviare on the house, and we must have eaten £100 worth during our five-day stay.

In search of the seamier side of Ankara, a couple of us wandered into the rougher quarter of the town, and found that we were being followed by two men in macs. But we were not worried: detectives are internationally recognizable. We went up to them, passed round cigarettes, and invited them to join us. They were friendly enough, but regretted that they were on duty. They had been ordered to protect us, and although they had to follow us, they could not join our party. Sport is heavily restricted when you've a couple of bloodhounds at your heels, so we returned to our hotel and went to bed.

Turkey's population is a confused mixture of Turks, Greeks, Armenians and Jews, with the non-Turks claiming a pretty raw deal. I met a Greek secretary: a tall, handsome girl, past the first youthful bloom, but retaining a magnificent figure. She was paid a fraction of what a Turk would have received for the job, and was unmarried. Some few years earlier the Turks had imposed a capital levy on non-Turks, and this had swallowed her dowry and had landed her brothers in debt. Without a dowry not even the most humble labourer would marry her, she said.

We had one evening out together, and a couple of letters followed me to England. They put the wind up me, for just then the *News of the World* came out with the story of a Greek girl who had followed a Naval officer to England and had involved him in an embarrassing court case. These girls take a quiet night out too seriously. And it *was* a quiet night out, for she had to catch the last ferry home, and to be caught even kissing in public in Turkey meant "disgrace" for the girl.

Unfortunately, despite the unarguable superiority of the Meteor, Gloster's were unable to unlock America's strangle-hold on Turkey. We sold no Meteors to the Turks—although we did achieve one useful purpose in terms of Allied defence strength. Our displays made the Turks dissatisfied with American war-surplus, piston-engined 'planes, and helped to get them American jets far earlier than they might otherwise have been offered to them.

6
ANOTHER WORLD RECORD

It had not been planned to display the Meteor in Greece. Athens was scheduled only as a refuelling point between Turkey and Rome. But we reckoned without our Greek agent, Alex.

Alex was a fabulous character. About six feet four inches tall, slim and hawk-like, his high-pressure tactics would have shamed the toughest American super-salesman. He decorated his rapid eloquence with purposeful gesticulations, and emphasized points by jabbing his fingers into your chest and arms. Alex was an operator on the grand scale. As Vickers' agent he was reputed to have sold the Greeks a cruiser, some destroyers and a submarine, and was determined that the Greek government should purchase Meteors. The poverty of the Greek exchequer did not worry him: he had some ingenious plan whereby America would foot the bill.

I explained to Alex, en route to Turkey, that I had no authority to display the Meteor at Athens. He replied: "Nonsense, I'll get their Majesties to say they'd like to see it." Sure enough, instructions came from England authorizing a display.

The show followed the familiar pattern. When it was over I taxied in front of the pavilion, shut down and climbed from the cockpit. Their Majesties wanted to see the Meteor close-up, and Stokes and I were to be their guides. An imposing collection of Royalty crossed the tarmac: King Paul—his six-feet-plus draped in immaculate khaki; the elegant Queen Frederika; ex-King Michael of Rumania with his fiancée, Princess Anne of Bourbon-Parma; and their mothers. We pointed out to them the finer points of the Meteor, and King Paul and ex-King Michael, both qualified pilots, proved most knowledgeable.

Then things became tricky. The diminutive Queen—like any wheedling wife—asked Paul in English if she could fly in the 'plane. "I want to go up, darling," she said.

No reply.

"I can, can't I, darling?"

Still no response.

"It will be all right, won't it, dear?"

She tugged at his tunic, and continued her pleading back to the pavilion. It was obvious that the King was not enjoying the situation, but he said nothing. Then the Queen turned to Stokes. "These engines and aeroplanes are safe, aren't they?"

Quick to defend our products, he replied: "Oh, yes, Madam. Safe as houses."

"You'd let your wife fly in it, then, wouldn't you?"

"Yes, indeed," said Stokes. "My wife often flies."

Triumphantly the Queen turned to her husband. "There you are! You see, there's no reason why I shouldn't fly in it, is there?"

"No, I suppose not," the King admitted.

The Queen turned to me. "You let your wife fly, don't you, Squadron Leader Waterton?"

"I'm not blessed with a wife. Your Majesty," I replied.

"Ah, but if you had a wife you'd let her fly, wouldn't you?" she persisted.

I don't know whether I unconsciously sympathized with the King, or whether my oft-misunderstood humour got the better of me, but I replied: "I would, indeed, Your Majesty. And if she was like some I've seen in my time, I might be inclined to leave her to her parachute at a great height."

The Queen snapped back: "That's very naughty of you," but the King's laughter, which echoed through the pavilion, acquitted me of *lèse-majesté*.

At that point we were joined by the American Army in the form of Lieutenant-General James Van Fleet, who was commanding the huge American mission to Greece. There was civil war at the time. I was reminded of an Italian friend's description of the type. "They are very kind, and work hard—but, oh! so clumsy." Van Fleet was a tough, flat-faced man of medium size, dressed in an open khaki shirt which revealed a grubby vest and a hairy chest, with his slacks tucked into his boots and his cap on the back of his head. He breezed into sight, roaring. "Where's Mike? I wanna see Mike."

Mike turned out to be ex-King Michael. The General grabbed him by the hand, slapped him on the shoulder and started a loud, one-sided conversation.

Someone asked: "What's that you've got, General?" pointing to a metallic ladder hanging down his chest. "Shootin' medal," said Van Fleet—and explained at some length how he came to win it. Then he took his leave with: "Well, I gotta go now. So long, folks." And off he went. The expressions on the faces of the assembled Royalty were like something from a Bateman cartoon.

This brought an end to the party, but as we started preparing the aircraft for its departure next morning, I found myself involved with another lady. She was wearing a light silk dress and a large, floppy hat, and had a beautiful, slim figure. She asked, in cultured English: "Are you the pilot of that aeroplane?"

"I am."

"Well, I want to go up in it."

"How nice," I replied. "And who might you be?"

"I'm Dorothy Norton. My old man's the Ambassador here."

"Oh! . . . Well, Lady Norton, I'd be honoured to carry you. . . . Tell me, what does His Excellency say?"

"Doesn't matter. I do things first, and talk about them afterwards—if at all."

"I'm awfully sorry, Lady Norton, but really it's quite hopeless. My general manager is very much against ladies in the company's aircraft. I've no authority to take you, much as I'd like to . . . . Anyway, we haven't much fuel."

"I'll fix that," she replied.

I played the ace. "There's one other point. We've just turned down the Queen. Surely there'd be diplomatic hell to pay if she heard you'd been in the Meteor when she hadn't flown."

Lady Norton gave up. She was quite crestfallen, and I've always felt badly about not giving her a ride.

It was true that I was worried about fuel, for Shell's advance dumps only allowed for scheduled arrangements. The problem cropped up again at Rome, where we had budgeted for four flights before moving on to Marseilles. An all-out display over the airfield was to be followed by a beat-up of the city, then two flights with Italian officers aboard.

We timed the displays for mid-morning, before the June heat of Rome caused the air to become excessively bumpy, but at the appointed time no Italian Air Force officers were to be seen. Sir Victor Mallet, the British Ambassador, told me to start the show without them. We had come to Rome in the hope of selling Meteors to the Italians, and I thought it a bit odd to give the display without the customers being present. But His Excellency was the King's representative, so off I went.

The show pleased the Ambassador, and as he moved away at its end a bus rolled into sight containing the Italians. I explained to them that the display over the airfield had taken place, and that shortage of fuel permitted only three more flights. Either I would do a second solo beat-up of the airfield, then take up two officers, or I would omit the solo flight and give three officers flights.

Understandably, this did not please the Italians, who were reluctant to miss anything that promised to be exciting, but I could not give way if I was to retain sufficient fuel for the flight back to Marseilles. I told them that I would take-off with one of their boys for a flight over Rome, and while I was in the air they could make up their minds about the remaining two flights.

My passenger handled the controls for a while, sampling the 'plane, then we "buzzed" Rome. I was told later that I had all but chipped paint from the dome of St. Peter's, and a Rolls-Royce chap who was friendly with a Cardinal learned that the Pope had been disturbed— then interested—as the Meteor screamed over the Vatican.

When I landed, the Italians put forward a compromise plan: to give three officers rides *and* beat-up the aerodrome by carrying one of them during the beat-up. I told them that I did not carry passengers on beat-ups—especially with a second stick in the rear cockpit. But they persisted, and middle-aged, grey-haired General Napoli announced: "I will go up with you."

"Fair enough," I said. "I'll make an exception this time. But I warn you that you may not like it."

"I'll be all right," he replied.

So we took off. I held the 'plane down until it reached the end of the runway, roared upwards in a vertical climb, then howled back to earth in a series of aileron turns. I climbed again, and let the General sample her for himself. Taking over from him again, I shrieked across the airfield at more than 600 miles an hour, some 10 feet above the ground. I glanced round and saw the General's knuckles all but burst - ing through the skin as he gripped the cockpit combing. I put the Meteor through the book—fast and slow, high and low, all ways up. I had just crossed the airfield inverted at about 30 feet, when a plaintive voice came over the intercom: "Please go back now. Please—we go back."

General Napoli climbed out, very green, and grinning feebly. He muttered: "Excuse me, please," and tottered behind a shed. A few minutes later he returned, smiling sheepishly from ear to ear. He clapped me on the back, and received congratulations from his brother officers.

No one will convince me, after that display, that the Italians, individually, lack guts. I wouldn't have touched the proposition for a hundred pounds: to be a passenger during an aerobatic display in a super-hot aircraft piloted by a man I'd never seen before—not likely!

On June 13, we flew to Marseilles, then Paris, a trip highlighted by bad weather forcing me down to 300 feet, losing my way, and eventually touching down with only 60 gallons of fuel (10 minutes of flying) left in my tanks. After seven displays in France, during which I carried Ministers of State and the top Generals, we returned to England.

It had been a fascinating month. The Meteor had behaved splendidly, covering some 5,000 miles without so much as a murmur of protest from her engines, or a rattle of disapproval from her airframe. I felt afterwards that the Mark VII was essentially my aircraft.

Unfortunately, our visits to Greece and Italy were as abortive as the trip to Turkey. We had never been optimistic about selling to the Greeks: the idea was our agent's. We knew that America had effectively ousted Britain from this traditional sphere of influence, and sure enough, despite the Meteor's brilliantly effective performance, the Greek Spitfires were eventually replaced by American aircraft.

Italy preferred—for some reason—the Vampire to the Meteor, but she, too, fell under the American spell. Today she is building American Sabres, and one or two trainers from designs of her own.

The French, however, did buy Meteors—though the number was small. They wanted them for experimental and training purposes, but later bought several squadrons of the night-fighter version.

The price of a Meteor varied between £30,000 and £35,000, depending upon the version, the equipment fitted and the quantity ordered, and the expenses for the trip—shared between Gloster's and

Rolls-Royce—were around £5,000. But the money was not wasted. Apart from "showing the flag", it taught us a great deal about long-range jet operations, for it was the first long-distance flight made in a jet aircraft, and it pinpointed many problems of navigation, flying control and meteorology—and showed the way to their solution.

Towards the end of the year it was decided to attempt another world record: for 100 kilometres (62.5 miles) over a closed circuit (using the same point for both starting and finishing).

Ideally the course would have been a deep, egg-shaped loop, with long, straight sides and the gentlest of turns at its end—for the tighter the turn the greater the speed-reducing drag. But the ideal was not for us.

The firm wanted to break the record, but were not prepared to spend much money in the process. Observers' posts, field transport and housing would have been needed if the perfect course was laid out. Instead, I had the job of finding a local course, its perimeter governed by handy landmarks, and the starting and finishing point the control tower of our aerodrome at Moreton Valence. It was no easy task, for the aerodrome lies almost at the level of the River Severn, with the Cotswolds to the east and north-east, the Welsh hills to the west, and odd pimples seven or eight hundred feet high to the north.

Limiting my choice further was the need for observers to be able to get to the turning points in a hurry when the weather favoured the attempt, and the necessity for them to be on the telephone. Even more: they had to be points on which I could line up with ease and accuracy—landmarks quickly recognizable irrespective of poor visibility. And the whole lot had to be exactly 62.5 miles round, from start to finish.

In the end I settled on the south bridge over the Avon at Evesham as my turning point to the left, then a short run to Defford aerodrome's control tower, an easier turn to the left for the run to a small railway halt on the Gloucester-Ross line, and so back to Moreton Valence.

With the county surveyor I plotted the exact turning points, and the distances between them were measured to four decimal points. The result was a rough parallelogram with only one turn sharper than ninety degrees. It was by no means ideal, but the best that could be done on our tight budget and the strictly imposed limitations.

I was to use a new production Meteor IV fighter. Since these were made on order for the Ministry of Supply, there was the ironic business of having to hire one back. It cost about £200 per flying hour when insurance and other charges were included.

I was not enthusiastic about the 'plane. The Mark IV's engines tended to resonate and rumble when they were started up. It could be corrected by manipulating the fuel cocks, but the R.A.F. pilots did not always do so. In service, engines were being ruined wholesale, so Rolls-Royce modified the fuel system to prevent both this and over-speeding at high altitudes. Logically, with the fuel delivery cut, engine r.p.m. dropped at low altitude and speeds were lowered. I calculated that to obtain an average speed of from 565 to 572 miles an hour, which was possible (the record at that time stood at 497 m.p.h.) I should need every fraction of the Meteor's top speed of 585 m.p.h., but the modifications had reduced this by about 15 to 20 m.p.h.

I complained and fumed, but to no avail. The pumps had been altered at the Rolls' factory, and that was that—even though the best possible results could only be obtained if the fuel delivery was increased to its previous level.

As if the odds were not enough, the publicity department added another hazard: one that all but cost me my neck.

The Press were invited to witness the attempt—and I was furious. I pointed out that newspaper reports about a *failure* to achieve a new record would hardly have prestige value, and that it would be better to break the record and then trumpet. But my arguments cut no ice. The newspapers were asked to come along, and this naturally involved the fixing of a specific date for the record bid—irrespective of weather conditions. I was aghast. It seemed little short of certifiable lunacy.

Sure enough, the weather on February 4, 1948, was appalling. A gale, with gusts that reached 75 miles an hour, was blowing, and scud raced across the sky at under 1,000 feet. I refused, bluntly, to take up the Meteor until the weather eased. Certain of the seventy-odd Pressmen crowded in the hangars and on the control tower did not conceal their annoyance, and the firm's publicity boys closed in on me. There was no direct order, but I had to put up with this sort of thing:

"Of course, you mustn't do anything silly, old boy. We wouldn't want that. It's entirely your decision. But this doesn't look good to the Press. Er . . . can't we do something?" The "we" meant me, and the "something" meant "have ago". Their ceaseless nagging became so irritating, and the atmosphere so unpleasant, that I finally weakened and said I'd go up.

I tied myself into the Meteor (VT-103), the control points were alerted, and the record attempt was on. As I taxied out the wind hammered at the controls with such force that it took two hands to stop the stick from banging from side to side. I literally stood on the rudder pedals to negate the wind's pressure on the rudder, and I roundly cursed the blithering idiots who could blithely bully others into risking their lives while they themselves kept both feet firmly on the tarmac.

At take-off the aircraft protested like a living thing, and I had to fight to control her, like a jockey trying desperately to master a fear-maddened horse. The gale was coming up the Bristol Channel from the south-west, and I flew into the teeth of it as I headed down the Severn. South of Sharpness, keeping just below the 1,000-foot cloud base, I turned to starboard to line up, the Sharpness docks and the Gloucester Canal pointing the way as I opened the engines.

As we raced faster and faster through the gale, the Meteor bucked and writhed, and I felt as though I was riding some fire-breathing monster from Hell. I kept just under the cloud until I was just short of the starting line, then went into a shallow dive to add a bit of speed as I crossed the starting line below the 300-feet limit. I hung on grimly as the airspeed indicator rose above the 510-knot mark, and for the thousandth time cursed the slowness of this modified Mark IV.

As I roared across the aerodrome it happened.

I was flying level at about 200 feet when a particularly violent gust of wind hit the Meteor and sent her into a shallow, yet determined, dive. Automatically I pulled the stick. It came back easily—too damned easily, for at more than 600 miles an hour the Meteor's elevator control was usually incredibly heavy: almost solid. But I was able to pull the stick right back—and the nose of the 'plane kept falling. Seven tons of Meteor, with me inside, hustled at the speed of a .45 bullet straight towards the experimental flight shed a few feet below.

It was a moment that no amount of writing can convey. For how can one describe utter and absolute terror? That moment of time—too short to record, yet as endless as time itself—when you are no longer a man, but an incarnation of fear itself?

I cannot tell why I did what followed, for it was against all good flying practice. But there was nothing left to do—and no time in which to do it. Possibly it was some remaining vestige of the instinct of self-preservation, maybe luck was with me—perhaps it was a combination of both those things. Anyway, I chopped the throttles, wound back the trim wheel, and rocked the stick forward and back. With a jerk the Meteor came under control, and I got her nose up. I think I could have touched the roofs of the hangars as I screamed over them. I was sweating and numbed with fear, but before a full physical reaction could set in I had my hands full again.

The Meteor seemed to have gone mad. She tried to loop one moment and dive the next, and this took a lot of controlling at over 550 miles an hour. I managed to land her safely, soaked with sweat, yet shivering with fright—and trying to show neither. The gentlemen of the Press did not realize how closely they had come to some really dramatic headlines.

Nothing was found to be wrong with VT-103, but I thought things over and came to a conclusion which I still believe to be the correct one. In the conditions, my aircraft should not have reached more than .8 of the speed of sound. In fact, however, she must have reached nearer Mach .9, and I realized that she had behaved exactly like any Meteor that exceeds its normal warnings of compressibility—control difficulties approaching the speed of sound. The change of trim and the dropping of the nose supported this theory, and the trouble was due to either a drop in the following wind, or a gust which affected the airflow over the 'plane. In either case a sudden increase in local Mach number seemed to have occurred.

I am convinced that this accounts for untold accidents to jet pilots flying fast, low down, in high winds and (or) gusty weather, and if my near-fatal record attempt served no other purpose, my report on it may have made some small contribution to our knowledge of jet flying.

I tried again on the evening of February 6. This time things went like clockwork, despite two unfortunate disadvantages: the slowness of my modified aircraft, and the fact that in the gathering dusk my path was slightly out, adding a couple of extra miles to the circuit. It created a new record—542.9 m.p.h. This was 46 miles an hour better than the old one, but a poor substitute for the 570 for which I'd hoped.

A few weeks later, Supermarine's raised the record to 560.6 m.p.h., but for financial reasons we did not relieve them of it—as we could have done. At the time, Supermarine's were busy trying to persuade people of their aircraft's superiority over the Meteor, and they finally sold some to the Navy and to Pakistan. But I hazard a guess that the Attacker-Meteor ratio of sales and production was twenty-to-one in Gloster's favour.

7

PEOPLE AND METHODS

In 1947 when "cold war" referred to nothing more sinister than commercial rivalry between ice-cream manufacturers, two of Russia's top designers paid a visit to Gloster's. The British Government had already given the Russians Derwent and Nene jet engines, and there was now talk of selling them Meteors. As it transpired, none was ever sold to the Soviets—although the Government did not actively discourage the idea. Perhaps, in those days, the Russians shared their secrets with us, too. I don't know.

Our Russian guests, accompanied by a beefy interpreter from the Embassy, were leading airframe designer Major General Artem Mikoyan, and Mr. Klemov, one of Russia's top engine designers and a Stalin Prize winner. Both were around fifty; Klemov slim and bespectacled, Mikoyan dark and slight.

Somehow, the two Gloster executives detailed to meet the Russians succeeded in missing them, and a wit suggested that our people had been foxed by the absence of fur hats, beards and snow-capped boots.

Anyway, the Russians suddenly appeared in the staff canteen with a works' driver, looking lost and baffled. I recognized them, organized coffee, and explained through the interpreter that our executives had probably missed them at the station. They laughed and said: "So it even happens in England."

I was invited by the firm to join the party for the evening, and we dined at Cheltenham's old Thurlestaine Hall Hotel—a converted country mansion that retained much of its original elegance. Our visitors were suitably impressed, although one of them, despite his fifty-odd years, remarked: "We seem to be the youngest people here." For a stranger to the town it was a pretty shrewd summing-up of Cheltenham's hotel society.

We managed to obtain vodka for our guests, but they did not live up to the popular conception of Russians at a party swallowing raw spirits by the pint, and drank fairly moderately. They said they had never been out of Russia before and knew no English, and displayed an intense interest in everything they saw, from the heating system to the cut of our clothes.

They spoke but little of aircraft. Having seen the excellent Yak fighter, and the PE.2 bomber, I brought them up in conversation. The Russians were polite, but gave nothing away. By the time we repaired to the bar, Mikoyan and Klemov were talking a fair bit of English. Then they asked to be taken to a cinema, and although the film may have improved still further their grasp of the language, it certainly spoiled our chances of learning Russian.

We did not sell them any Meteors. We padded the 'plane's performance figures, but they laughed amiably and said: "Come off it." They are probably still laughing, for a year later Mikoyan's brilliant Mig-15 took the air, while at the time of writing the R.A.F. are still equipped, in the main, with the inferior Meteor.

My years at Gloster's brought me into contact with a wide diversity of people, and not long after the visit of the representatives of Communist Russia I found myself shaking hands with Prince Bernhardt of the Netherlands.

He came to see the Mark VII two-seater, and arrived piloting his own Dakota. Again the official welcoming committee failed to turn up in time, and it was left to me to receive the royal party. After the usual introductions, the Prince said he'd like to go up in the Meteor right away instead of waiting.

As he adjusted his too-tight parachute harness, formality disappeared, and we were put at our ease by the Prince's fluent and colloquial English.

The Prince was an extremely competent pilot, and my presence in the 'plane was quite unnecessary as he put the Meteor through her paces.

The Dutch Navy was interested in Hawker's Fury, and one was flown in for Bernhardt to try. There was an anxious moment when he radioed that his engine was misbehaving, but he landed satisfactorily except for going slightly off the runway. An oil gauge had packed up, and the Prince rightly thought it signalled engine trouble. Fortunately, however, the oil pressure was all right: only the gauge system had failed. He did not seem unduly concerned about the incident—which is more than can be said of the rest of us.

In July, 1948, I took up my second royal passenger: the Shah of Persia. This was at the Bristol Aeroplane Company's aerodrome at Filton, where a Bristol freighter and a helicopter were also to be displayed to the Shah. I flew in from Gloster's to open the show with a beat-up of the aerodrome, and as I landed a freighter took-off with the Shah abroad. I was told that he wanted to fly in the Meteor when he came down, but Bristol's, hoping to keep to a tight schedule, were not happy about it, for our salesmen had laid it on during the Meteor's display. "Make it fast," they told me. "Five minutes only."

The Shah's English was not a hundred per cent. reliable and I was worried about passing instructions to him over the intercom, but we got by on a mixture of French and English.

He, too, was an excellent pilot—he flew his own Hurricane—and

I had no fears about handing over the controls. A view of the local countryside was too tame for him so we carried out aerobatics and saw 600 miles an hour in a dive. By this time Bristol's programme was wildly cockeyed, but my hints about getting back had no effect. In the end I had to take a firm line, and the Shah handed over with: "You have command now." He was delighted with the flight and obviously reluctant to exchange the Meteor for a helicopter.

Lesser V.I.P.'s came thick and fast, and when I flew at Farnborough that year I was not at my best. Constant beatings-up had caused me to grow stale.

At about that time I delivered a Meteor IV to the French and met a number of their military and civil test pilots. Many were also engineers and struck me as being a much more highly qualified body than their British counterparts. They were also better paid. I returned to England by Air France and was given handsome V.I.P. treatment. At the take-off I was invited to go for'ard and see the captain. It was a blistering day, and he sat in his shirt-sleeves. His engineer wore a singlet. To my surprise, they did not climb when they left the ground, but headed north at tree-top level.

The explanation was simple: "We are looking for the *Tour de France* which is approaching Paris." Schedules were forgotten until we found the cyclists and made a number of low circles around their colourful procession. Finally, the captain said: "Now we go to London. Want to fly it?" I nodded, and took over as far as Tunbridge Wells. I told the captain I thought the Languedoc was a good, solid aircraft. He agreed, and demonstrated its toughness by deliberately "bouncing" it on landing. We arrived at London Airport twelve minutes late—but we had seen the *Tour de France,* and all twenty-three passengers seemed happy enough.

For some reason of which I am ignorant, the firm was reluctant to allow women to fly in the company's aircraft. This ban was a sore point among the Gloster pilots, for those at Hawker's (a sister firm) flew their wives everywhere.

While in France on my way home from Turkey there had been a rumour that Jacqueline Auriol, daughter of the French President, and a noted aviator, was to fly the Mark VII, but nothing came of it. Then Diana Walker, an ex-A.T.A. pilot, widow of Wing Commander Derek Walker, and daughter of racing-motorist Wolf Barnato, wanted to take from America Jacqueline Cochran's women's speed records. Someone in the London office thought the idea a good one, and I was told to be prepared for her arrival and instruction. In a Meteor she could easily have taken the records, and although she could not obtain a 'plane from the R.A.F., Gloster's privately owned Mark VII was available—except for the firm's disapproval. The go-ahead came from head office in London, and I got as far as instructing her in the Meteor's controls and systems. Then my instructions were changed. I was ordered to choke her off, and she never actually flew. A pity, since it would have been of prestige value to both Gloster's and Britain—and she did receive initial encouragement.

Later another female entered the scene, when the Press splashed her as the girl who wanted to be the first British woman and mother to break the sound barrier. The Javelin had taken the air by this time, and "Jackie" Moggridge came to see me, convinced that she was just what Gloster's needed to put the Javelin on the map. Her introduction was through a friend of hers—one of my pilots who thought I and the firm would be helpful. I suggested that he should hire a Meteor from the Ministry of Supply and give her twenty hours' flying instruction, then hire a Sabre in which to go through Mach 1. He didn't think it very funny: it's odd how humourless people become when you suggest they spend their own money.

Portuguese . . . Belgians . . . Syrians . . . Swiss . . . Israelis . . . Venezualians . . . Pakistanis . . . Egyptians . . . Spaniards . . . Swedes . . . Canadians . . . Danes . . . Norwegians. They all came to fly in the Mark VII. When the Javelin took the air, many more people were interested and came to look, but only the Americans were allowed to fly.

In October, 1948, a surprising interest in the Meteor was shown by

B.E.A., and their then Chief Executive, Peter Masefield, came to Moreton Valence with his number two, Peter Brooks. Both flew in the Mark VII and spent a great deal of time discussing jet operations. They showed particular interest in my long overseas trips, and we had detailed talks about the problems of aerodrome control and procedure, navigation, radio communications, fuel consumption, range and endurance.

A number of hush-hush meetings followed their visit, all wrapped in deep secrecy. It was finally revealed that B.E.A. were to receive two Meteors for use as mail 'planes between London and Paris. Route operations, traffic patterns and jet problems were to be studied—but the G.P.O. and the Government would not give the scheme their blessing, and it was scrapped. As with helicopters, Masefield was way ahead of the field, but "the powers" failed to support him.

Later, when the de Havilland Comet came along, we tried to interest B.O.A.C. in the idea of giving their pilots jet training in Meteor VII's. From our own experiences, I knew many of the problems jet airliners would run into, and I wrote about them in the Hawker-Siddeley *Review*. It did not make me very popular with de Havilland's. My article said: "Safe, high-altitude passenger lines are five or ten years away," and referred to such matters as high-altitude turbulence, thunder-clouds above 40,000 feet (the airlines were advertising: "Fly with us over the weather at 20,000 feet"), pressurization problems, rapid temperature changes and electrical troubles. Unfortunately, the Comet had to do it the hard way.

Had we encountered cabin fatigue we might have had advance warning of the trouble the Comet was to run into, but our pressurization was lower than theirs by almost two-thirds. Even so, Meteor canopies often burst, with sometimes disastrous results to the R.A.F. There were also lesser cases of over-pressurization, and I remember flying back to the works a grounded R.A.F. Meteor which had part of its fuselage wall blown out like a bunion.

The cause of these troubles sometimes verged on the comic. One

R.A.F. Meteor Squadron returned from Belgium with an epidemic of pressurization problems. The cause, incredibly, was due to chocolate, at that time rationed in Britain but not in Belgium.

In a pressurized 'plane, air is fed into the cockpit through pipes, and can only escape through the pressurizing valve. In their wisdom, the designers had placed this in the cockpit floor. Since there was a flow of air towards it, the valve sucked up the wrappers discarded by the chocolate-chewing pilots as they flew along, together with dirt from the floor. Eventually Gloster's were contracted to move the valve to the cockpit's vertical wall.

In October, '48—the same month in which we were visited by B.E.A.—Charlie Gardner, then with the B.B.C., broadcast from the Mark VII in flight. He talked through the normal R/T set, his voice being picked up in the control tower. He grew wildly excited as we built up speed, and I was deafened as he screamed his drama to the listening millions. Charles was the first of a long procession of B.B.C. types who turned up regularly every few months for repeat performances. I think it became an amiable skylark for them rather than anything else. Busy with other things, I came to give the job of flying the B.B.C. man to a new boy. The old routine was repeated—and both B.B.C. chap and pilot thought they were hellish fellows.

At about this time *The Aeroplane* sponsored a stunt to show how closely cities could be linked by the combined use of helicopters and jets. From St. Paul's, London, a letter would be taken to the Mayor of Paris at Les Invalides within an hour. A Sycamore helicopter would take the letter from St. Paul's to Biggin Hill, I would fly it to Orly airport, and a second helicopter would carry the missive on to Les Invalides.

The publicity got under way, and the experiment was scheduled to take place on the morning of October 30. But visibility was poor on that day. When Biggin Hill was free of fog, there was fog over Orly, or St. Paul's, or Les Invalides—or both. Since mine was the trickiest flight, I had the deciding voice. Such were the pleading looks on the organizers' faces, I agreed—reluctantly—to have a go at noon, although the Paris weather was far from good.

After a frightening moment when it hit some wires at St. Paul's, the first helicopter collected the letter, and the curtain was up. I followed the helicopter's progress over my radio, and the messages to the Biggin Hill watch-tower were relayed, in turn, over the B.B.C. When the helicopter was about three minutes away I started up. The Sycamore landed a few feet off, and the letter was rushed over to Eric Greenwood—Gloster's publicity man—in my rear cockpit. The lid was slammed home, and we were off.

Navigation was spot on, and we made good time. In five minutes we were over the Channel, and twelve minutes out of Biggin saw us past the French coast at 15,000 feet. This side of Paris, however, our near-600 m.p.h. progress was slowed by poor visibility. I went down to 800 feet, able to see no more than two miles ahead. I hoped to God that another 'plane wouldn't get in the way, for we were still topping 540. I recognized Versailles—and then we were lost.

Visibility was reduced to a mile—some seven seconds' warning of obstructions. Not much when you're down to between 300 and 500 feet. At this point the intercom packed up, so that if Greenwood spotted any familiar landmark he could not tell me. I missed Orly, but recognized a wood just east of the Seine which swings south-east of Paris and is to the east of Orly. I started a "square search" back and forth across the river. At last I found Orly and went in from the north. The letter was picked up by the second helicopter and rushed off to Les Invalides.

We had taken nearly twenty-seven minutes, four of them looking for Orly. Not good. But the helicopters did better than planned, and the letter made its St. Paul's/Invalides journey in a total of forty-seven minutes. I believe that is still the fastest postal delivery between the two capitals.

We received commemorative medals from the Aero Club de France at a banquet they gave. Later in the evening, while jocularly testing the medal by biting it, I broke a tooth, adding a second souvenir to the occasion.

Although an aircraft factory is a complex organization, it has only grown to be so in the past fifteen years. There was a certain change between the wars in the building of civilian 'planes, but the construction of military aircraft varied little. As late as 1940 personnel still consisted mainly of fitters, riggers, welders and sheet metal workers, plus a few machinists, painters and fabric workers.

The first major change was an emphasis on metal rather than fabric, but it was not until near the beginning of the war that there was any fundamentally new approach to the construction of aircraft. Inspired by demands for better speeds and performances, hydraulics crept in to raise and lower undercarriages and flaps, and to operate turrets and bomb doors. Pneumatics made their bow. Then the most important change of all: the introduction of electrics, virtually non-existent in British pre-war military aircraft. Bombing and navigation equipment, gun and bomb sights, together with radio and radar, made enormous electrical demands. The basic flying machine became a complicated mass of machinery, and the handful of workers grew into a small army of technicians and scientists.

The jet was hailed as the "great simplifier". And so it was—until its own particular, hitherto unknown, problems demanded further innovations. The old controls and facilities were inadequate for the new speeds and ceilings. Hydraulic flying controls had to be introduced, together with trimming and flying tailplanes, cockpit heating and pressurization, new cooling systems for electrics—and a score of other new introductions.

The small army became a large one and, logically, the complex monster of aeroplane manufacture now demanded long planning and detailed organization. But in many cases this was lacking, and the ironic situation arose of the jet, which Britain pioneered, becoming a Frankenstein's monster. Gloster's gave Britain the Meteor, but the industry was not ready for the further development of jet aircraft. Consequently, when the Americans produced something better than the Meteor, the industry could only limp in pursuit.

In retrospect it seems remarkable that Gloster's managed to produce the Meteor, because they were as unprepared as any for the involved ramifications of high speed jet flight. Yet, in the immediate post-war years, we led the world, despite the fact that we lacked comprehensive research facilities, involving electronic computers, high speed wind tunnels, and the mass of other expensive and essential paraphenalia needed in this faster than sound era.

However, the organization was still pretty complex, and resembled a small, self-contained town. It employed 12,000 people during the war, 3,500 when I joined it, and 6,000 when the Meteor was at the peak of production. There was a "welfare state" to look after the workers: canteens, surgeries, welfare and labour relations staff, and a transport organization. Then came police, maintenance engineers and a fire brigade.

There were two inspection organizations: the firm's and the Government's, and other resident "watchdogs" from the Ministry of Supply. A contracts department negotiated Government and foreign orders, and a finance department kept its eyes on the cash.

The senior executives were the general manager, in charge of the whole show; the works manager, responsible for production and maintenance; the secretary, who looked after finances; a couple of professional directors; and the chief designer. Then came the departmental heads: production manager, service manager, two chief inspectors, resident technical officer, contracts manager, security officer, assistant chief designer, drawing office manager, sales manager, and the publicity boss.

And so on down the line to the men who handled the tinsnips and spanners.

Even when the Meteor was past its prime, Gloster's flourished, for they were firmly supported by Government contracts—and what other 'planes were there to be ordered in the Meteor's place?

There was a time when 'planes were produced on the open market: a firm relying on the quality of its products to guarantee sales. Today, most free enterprise is dead. Rather than put faith in its own designs and build on the basis of orders following results, the industry prefers to stick to the safe subsidies provided by Government contracts. (Even the de Havilland Comet, allegedly a private enterprise venture in design and building, proved, finally, to have been heavily subsidized.)

One firm in our group—Avro's of Manchester—did try their hand at a post-war civil airliner: the Tudor, based upon Lincoln bomber design. It was not a conspicuous success, and maybe that discouraged the others.

The Services have received a number of dud designs since the war. Men have died in them. Yet the aircraft companies flourish, for they have played safe carrying out Government contracts when a little speculative free enterprise and competition might have produced better, safer 'planes.

Three parties are concerned in this subsidized supply of military aircraft: the Service concerned, the Ministry of Supply, and the manufacturing firm.

The Service has a difficult job. Its chiefs must assess the potential threat to the country, and devise ways and means of dealing with it. At the same time it must be prepared to adopt an offensive rôle in time of war. Yet the Service chiefs are hampered by a tight budget and must plan, not for today, but for ten or fifteen years ahead. It is no easy task, but once they have evolved, with M.O.S. scientists, what they want, they issue their requirements. These state the desired speed, range, height, armament, rôle, etc. The Ministry put the requirements out to tender.

On receipt of the specification, a firm decides whether it is interested. Some only build bombers or transports, others concentrate on trainers, while there are those which touch nothing but fighters. If a firm is interested, its design staff's Project Section makes what is called a "design study". This is an illustrated brochure containing the firm's suggestions for turning the Service's needs into an aircraft: figures of estimated performance and drawings of the proposed 'plane.

From the design studies received, the Service and the Ministry usually select two firms to make prototype 'planes. Although the basic specifications are the same, designs used can be widely different—as with the Gloster Javelin and de Havilland 110. The reason for giving contracts to two firms is very practical: if one flops the other will probably come off.

Theoretically the two best proposals get the contracts, but in practice the Ministry of Supply farms out its orders so that there is always work for everyone in the industry.

The prototypes are, of course, hand-made, and although the practical thing would be to see what they are like before committing a firm to production, political events—and the desire to keep the firms busy—sometimes result in 'planes being ordered straight from the drawing board.

The firm is paid advance sums to buy materials and to tool-up. This involves the designing and building of the heavy, set-in-concrete jigs, and the design and manufacture of press and machine tools. This tooling-up is the biggest single expenditure in a new 'plane, and in the case of the Hawker Hunter was reputed to have cost £8,000,000.

The great day comes when the prototype is wheeled out for its ground and engine checks. It is an even greater day when the 'plane commences taxiing trials. And the greatest day of all, of course, is when the prototype takes the air.

It has been made to measure. Very little of it will be common to any other 'plane. Every piece of equipment has been specially tailored and tested in relation to all the other items of equipment.

Wheels, brakes, tyres, hydraulic equipment, cockpit, engines, flaps, controls—all are new. Thousands of calculations have been made, and every possible complication supposedly allowed for. But there is still the possibility of something going wrong—some tiny error which will destroy years of work and millions of pounds worth of money. There is always a chance that the prototype will end up a blazing, twisted, useless wreck.

That's why they have test pilots.

Since an aircraft must both fly through the air and be controllable while so doing, its performance and control are the two principal things with which the prototype's test pilot is concerned. At the same time, and as a complementary job, he must also check the engineering: the various items of equipment assembled in the aircraft.

Consequently, the testing is broken down into several parts, the three main ones being performance, handling (the control of the 'plane as desired while it flies), and engineering. Handling is further sub-divided into qualitative and quantitative, the former being the pilot's opinion of flying qualities, the latter their physical measurement.

As with a car, there are certain fundamentals of "handling" behaviour which should be common to all aircraft, irrespective of size, age, design, make, modifications, or the inclusion of revolutionary ideas. When, for example, you take a car round a corner, you expect to place a moderate force on the wheel in order to negotiate the bend, then, when you release your hands, the wheel should centre itself and the car carry on in a straight line. An aeroplane ought to act in much the same way. If, after induced deviation, it does return to straight, level flight you have the ideal of "positive stability". If, however, when you take your hands from the stick in, let us say, a dive, the 'plane continues to dive—neither more nor less steeply—it is "neutrally stable". (For some purposes this condition is acceptable, although it means that the pilot must fly the 'plane all the time, and dare not take his hands away: a circumstance that does not help instrument flying.) If, on the other hand, the 'plane *steepens* its dive, you have "instability": a dangerous state of affairs since the 'plane can quickly get out of control and break itself up.

A further point to remember is that, unlike a car's one-dimensional movement, the 'plane's movements are three-dimensional—up and down (longitudinal), side to side in level turning (directional), and rolling (lateral).

A large part of flight testing is devoted to getting correct stability, a basic requirement without which no 'plane is of much use.

The next feature to be checked in the "handling" section is control, and this is an apparent paradox—to make certain that the 'plane is able to defy its stability. Yet it is not really such an absurdity, for just as you want your car to be able to negotiate bends, so the pilot does not always desire to fly in a straight line—although he wants his 'plane to return to the straight line when he takes his hands away, he expects to be able to change direction and height at will.

One way of achieving this is to build plenty of stability into a 'plane, then instal powerful controls to shift the aircraft despite its natural reluctance to depart from its flight path. This is the American method, and to a large extent was the German, too. The alternative is to have less stability and, as a result, less powerful controls since the forces to be overcome in altering the flight path will not be so great. British designers often favour this method, particularly for fighters, as it means that the 'planes need not be built so heavily, and therefore can be made lighter. On the other hand this method demands far more of the pilot.

Stability and control, it can be seen, are strongly linked, and with today's heights and speeds it is not easy to strike a balance between them. Both control and stability decrease with altitude, and near their ceilings aeroplanes feel as though they are precariously balanced on pinpoints. This is caused by the thinness of the air. At varying speeds airflow differs and also affects control and stability.

This work, which occupies so much of a test pilot's time, is by no means as exciting as popular belief would suppose—unless something goes radically wrong, when it is then up to the pilot, drawing upon his experience, and depending upon his reflexes, to save the day as best he can.

The pilot's qualitative observations during these tests are checked quantitatively to see how much his findings vary from the ideal. Measurements are taken to show how far, for example, the control column has to be moved, and how many pounds' force are needed to shift it in order to put 1, 2, 3, 4, 5 or more G on the 'plane—"G" being the pull of the earth's gravity, which is used as a measurement of acceleration. One G, for example, is what the plane weighs—say 10 tons; three G is therefore 30 tons—and so on. Checks are made to see how fast the 'plane will roll, how much rudder is needed to counter engine failure on one side—and so on. During most of my testing the pilot checked these things with spring balances and measuring sticks or tapes, recording his results on a knee pad. Later, automatic electronic devices—strain gauges and desyn recorders—were connected to various controls, and their readings photographed in the auto observer. For high speed aircraft these gadgets are essential—although, in my experience, the auto recorder's reliability was not to be taken as gospel.

The figures from these observations are then plotted on graphs and checked against ideals established from hard, practical experience accumulated over the years.

So much for "handling". Involved in "performance" flying is finding out what the new 'plane will do, and establishing the best way to use it in order to get the maximum speed, climb, etc. Again it involves long, painstaking, accurate—often boring—flying. The drag of the 'plane is measured by noting its speed at a given height against the engine power used, or by checking the time taken to accelerate and decelerate to and from given speeds. Alternatively, drag can be determined in glides at low speed—the height lost is measured against speed. To establish rates of climb, climbs are made at various speeds, steady or varying in the climb. Steady speed and power is maintained for a few thousand feet at various heights to obtain rates of climb at any given height. Here again, the automatic observers have taken over from the stop-watch and knee pad, but since instruments can give false readings of speed, height and Mach numbers, position errors are discovered by flying a new 'plane alongside one whose instruments are calibrated, and with radar-timed measurements from the ground.

Little of this figure collation is anything but dull in the ordinary run of things, and any careful, precise and accurate pilot can obtain the information if properly briefed. From the subsequent calculations, decisions can be arrived at to decide the best way of improving performance, control or stability. And when these modifications are introduced, the appropriate series of tests must be repeated, for if, let us say, a new tail, elevators or ailerons are fitted, more than that feature is affected. The entire stability of the 'plane may have been altered and the most unexpected features of the aircraft influenced.

While all this testing is going on, the pilot must watch the engineering—engines, electrics, fuel system, hydraulics, pneumatics, radio, instruments, and more besides. All are either new or are being used in a particular manner for the first time. So the pilot must keep his eyes, ears and nose open to spot faults when they reveal themselves: a fire in the air, fuel starvation, overheating, failure of hydraulics or electrics, and so on. These flaws are always cropping up, and their investigation and rectification keeps new 'planes on the ground for a large part of their early lives. As a result, it is rare for a British 'plane to fly more than a hundred hours in its first year. Normally about forty of these hours suffice to cover the handling and performance tests of a prototype, but modifications to engineering faults increase that figure many times over.

Yet, until these bugs are sorted out, the 'plane should not be put into production. Only when a 'plane is cleared by the experimental pilot is it fit for large-scale production. Often, however, for reasons of politics or expediency, 'planes are rushed out before they are fully up to scratch, and are issued to the Services with limitations and restrictions placed upon them in order to avoid disasters. This asks a lot of young squadron pilots, but such decisions seldom come from the manufacturer's flying men.

In any case, with the limited aircraft and time at his disposal, no test pilot can really cover every contingency. New troubles occur in service as 'planes deteriorate with age and hard use.

Hardly ever do 'planes meet the requirements of A.P.970—the Air Ministry design and test bible. In recent years only one 'plane is reputed to have done so: the Martin-Baker V fighter—and that was never ordered. Consequently, 'planes are a compromise, and even at their best are never what they should be, and by the time the R.A.F. receive the 'planes it is too late for anything to be done. The Ministry of Supply receives six copies of the test pilot's flight reports, and have their technical staff on the spot, yet the warnings in those reports often go unheeded, despite the test pilot being the only person who knows the full facts about the aeroplane, for in the air he learns of things that can never be discovered by the boffins on the ground. One of the most aggravating features of experimental flying today is the way in which boffins have risen to the fore and are now trying to tell experienced pilots how to fly aeroplanes.

So the chief test pilot's job involves not glamour, but precision flying in the air, and often heartbreak frustration on the ground. There is only "excitement" when something goes wrong, for anything can happen in a new aeroplane. . . .

In the middle of a transonic dive I've had tail-wagging and banging like a battery of machine-guns. . . . I've had ailerons and rudders take control and overbalance. . . I've known elevator control to reverse at low speed and at the stall. . . I've had complete longitudinal instability set in. . . . I've had flutter. These were the times when test flying isn't every pilot's meat. Only recently I was talking to one of the few test pilots who has made old bones, with many years experimental flying behind him, and said: "Despite power controls and all the rest, there's still a lot of flying by the seat of your pants." He agreed.

Like dogs, 'planes usually bark before they bite; like horses, they twitch before they kick or buck; like many boxers, they telegraph their punches. This is transmitted in many ways, and a test pilot must learn to recognize even subtle variations in feel, sound, vibration and smell. Not only recognize them, but must know how to act, for it is truer of experimental flying than of anything else to say that there are only two kinds of people: the quick and the dead.

8

EXPERIMENTS, TRIALS AND THE LATER METEORS

In May, 1947, carrying out trials on RA 421, a modified Meteor IV which was being prepared for cold weather tests in Canada, I made my nearest-ever approach to the speed of sound in a Meteor.

Normally the Mark IV ran into serious compressibility—control difficulties—at .8 of the speed of sound. At about .76 the nose would start to come up. This could be trimmed out by the elevator tabs which took the load off the control column. Above .78 the 'plane started to pitch and buck and the wings dropped to either port or starboard—or alternately to each. If the pilot persisted, he could get to a higher Mach number—.82 or .83—by diving at full throttle from a high altitude, but at this stage the Meteor would be shaking with fierce violence—the push force on the stick would change to a pull force, since the nose was inclined to drop rather than rise as it had been doing as speed built up.

On May 9 I took RA 421 to some 45,000 feet and put her into a full throttle dive. The air speed and Mach number needles crept steadily round their dials. They slowed up at about .78 to .84, hovered reluctantly, then, to my surprise, started to wind-up again. I pulled the stick back—but there was no proper response. The dive *steepened* and the nose failed to come up. To stop my spiral descent, I chopped the throttles. She slowed down, and I regained control at 25,000 feet.

This failure of a 'plane to respond to the controls when approaching Mach One provided one of the more dramatic moments of the film, "Sound Barrier". But when the nose goes down under compressibility, pushing the stick forward (as in the film) will not pick it up. If anything, it will worsen the position. The only really authentic part of the scene was the 'plane's shaking as it approached the speed of sound.

My speed in the dive turned out to be Mach .9—about 600 m.p.h. instead of the normal 530 maximum at that height—which was exceptionally good for a Meteor. I never claimed to be flying the 'plane during the fastest part of the dive, however: it was flying me.

Experiments were always under way at Gloster's, and a great deal of my work involved testing various modifications.

We tried out rain deflectors and clear-view paste to discover methods of enabling pilots to see in the rain, for raindrops splattering flat screens made them opaque. My experience with early U.S. Thunderbolts showed that in all but the worst conditions their "V" screens gave reasonable vision: the water flowed off in rivulets instead of splattering. The Americans still use such screens, but our long-haired boys say they are no good for aiming through. So R.A.F. pilots fly almost blind when it rains—which is not infrequently in Britain—and are unable to aim at anything.

Early on, jets' wheel brakes gave a lot of trouble, for the jets' clean lines, tricycle undercarriages and lack of propeller drag, encouraged them to go on moving. The old brakes weren't up to it. The fabric air bags we used blew up with the heat, the drums became distorted, linings faded, and friction heat often caused the brakes to catch fire. Day after day we made interminable take-offs and landings to check distances, temperatures and speeds, testing new equipment, until, finally, satisfactory brakes were evolved comprising twin bags and bi-metal drums. Later came hydraulic brakes, using flat, heavy copper plates. These worked admirably, but in time of war there was likely to be a shortage of copper, so special steels were evolved. Each stage of development involved hundreds of tests.

We tried new fuels, and new fuel venting systems to cope with different vapour pressures of paraffin and gasoline-based fuels. We dealt with cases of fuel tanks collapsing in dives and climbs because external and internal pressures did not equalize rapidly enough. We experimented, scores of times, with gun heating arrangements. We frequently flew at night to check reflections and cockpit lighting and heating.

Tail parachutes were streamed and trailed at various heights and speeds, while special equipment photographed them and measured the strain. This was for a Farnborough investigation into the behaviour of pilots' 'chutes at high speeds and altitudes. They were thought to blow out or collapse under shock loads, especially if they held moisture and froze on being released. We would climb to 40,000 feet, get settled, then move a switch which streamed the 'chute. Clear weather and correct winds were essential for this, since the tests were performed over the Bristol Channel, and we had to make certain that the metal fairings which fell when the parachute was released did not hit shipping. The 'plane then towed the 'chute back to Moreton Valence at low speed, where it was dropped off for examinations and further use. In case of trouble the 'chute could be jettisoned over the Severn. I made it a rule that the river route must be used to get to and from these trials, for a fatal accident could result if a stray parachute drifted over the windscreen of a moving land vehicle.

We flew with rockets and bombs, Meteors being tested to carry up to twenty-four 100 lb. rockets, or wingtip tanks and two 1,000 lb. bombs. The Meteor was a good load carrier. Although heavy, she had an excellent power-weight ratio, for while two 1,000 lb. bombs were only a seventh of her weight, they were a quarter of a Vampire's. As a result, her performance suffered less with a load than did a smaller aircraft's.

We flew on many trials for Rolls-Royce's Derwent engines. Additionally we tested instruments, radios and radio compasses. There were also tests to discover weak points in the Meteor. For these, batteries of recording instruments were connected to strain gauges. We would do a manoeuvre, then repeat it again and again, more and more violently, until the instruments told us there was nothing to worry about—or that it was time to pack up before something gave way.

I also handled and cleared the earliest Meteors fitted experimentally with ejector seats. They were installed in the ammunition bays of ancient hacks and necessitated a great open hole at the back of the cockpit. They were flown on behalf of the seat's manufacturer—and I discovered that Gloster's charged £15 per hour's flying for my services, plus travelling and subsistence allowances. If Gloster's had paid *me* at that handsome rate there would have been something to write home about. For my annual flying often totalled 400 hours: equivalent to £6,000 a year—plus travelling and cost of living.

We built two special experimental jets for the Navy: Mark III's fitted with the more powerful Derwent V engines. Constructed with the short take-off and landing distances of aircraft carriers in mind, they weighed only 12,000 pounds, and could get off in 185 yards on a calm day. They climbed like rockets and could land at 75 m.p.h. The Navy said they were the best jets that had ever done deck trials with, but the Meteor never went into sea service—although the Admiralty bought Mark VII trainers which were flown from land bases.

Other work involved cockpit heating and screen defrosting. Coming from a cold country, I suggested a method of coping with the latter which worked on Canadian car windscreens. The idea was to attach panels to the cockpit screen, inside which would be fitted electric wires to prevent icing. My proposal was used on a large scale, and although Gloster's patented it I cannot remember, as I live, anyone even thanking me for the suggestion.

While the Meteor was enjoying its world supremacy, it was in constant demand as a test vehicle for new experiments and engines, and we built three 'planes to take special engines made by Metropolitan-Vickers, Rolls-Royce and Armstrong Siddeley.

The first was RA 490, which initially carried 3,850-pound thrust Metro-Vickers Beryl engines. I flew it in September, '47. On its second flight I crashed it on the aerodrome on its belly at 200 m.p.h. when the hydraulics failed: fuel was short and the engines would not throttle back to idling, which accounted for the high speed. It was repaired and I flew it again a year later. The Ministry of Supply were satisfied with it and it was passed to the National Gas Turbine Establishment for further trials. It went to 50,000 feet and would reach 40,000 in less than two-thirds of the time taken by a normal Derwent-powered Meteor, despite being a couple of thousand pounds heavier.

The second special was RA 491 which packed a terrific punch in its twin 6,500-pound thrust Rolls-Royce Avons. Although Rolls-Royce installed the engines themselves at their aerodrome at Hucknall, Notts., it was still our responsibility to clear the 'plane as a flying machine. Rolls-Royce protested about this with something approaching a howl: apparently they didn't want any outsiders to get near their new engine. They said to me: "If you must fly it, just whip the 'plane around once at Hucknall, then we'll take it over." I replied that I had a job to do, that I intended to do it properly, in full accordance with our contract, and that I wanted RA 491 at our own aerodrome for ten days, to cover the contingencies of five hours' flying. I took the line that if Rolls-Royce wanted to do it their way, they should go to the M.O.S. and get the contract changed.

There was a high level rumpus. Rolls were overruled and I went to the Midlands to bring the 'plane to Brockworth. I took off from Rolls-Royce's rough grass field at a weight of more than 18,000 pounds: two tons more than the normal Meteor IV. The 'plane went well, and Rolls-Royce had her back, cleared, in less than my estimated time. She was extremely powerful and would get to 40,000 feet in under four minutes. A drawback, however, was that her weight needed a lot of stopping on landing.

Her trials, too, included an "incident". At a height of more than six miles, and still climbing at full power, the engine temperatures approached danger levels and I throttled back. To my dismay, the port throttle lever was jammed hard in the fully open position, and we continued to hurtle upwards, for one Avon had almost the power of two Derwents.

I used all the strength of my two hands on the throttle. Something gave, and the engine throttled back—although the lever remained horribly stiff. I left it alone in a "back" position, cutting the engine on touch-down. The alternative would have been to land on one engine—not to be recommended with a heavy 'plane I had flown only twice before.

Later, RA 491 suffered a rougher landing in the hands of Rolls-Royce. They repaired her, but complained of vibration. I checked and adjusted the aircraft in a further series of flight trials, involving "stick-jerking" flutter tests up to full speed. This meant hitting the stick and rudder bar hard and allowing each to settle back, flying hands and feet off the whole time. If it damped quickly one moved up to a higher speed and tried again. If the control was slow to stop moving or tended to keep moving one slowed down quickly, for flutter might be expected at higher speed and changes would have to be made before one proceeded. She was finally sold to the French who tested their own jet engines in her.

The third special: WA 820, with a Mark VIII tail, contained even larger engines: Armstrong Siddeley Sapphires, each with 7,600-pound thrust. I was abroad when she came along, and another pilot made the maiden flight. I was back in time for later trials—and was involved in a further spot of bother.

After a flight I landed her—but couldn't get her stopped. Although my brakes were fully on, I was still doing sixty, and there were only 400 yards of runway left. Beyond it were trees. In such circumstances experience and training take charge, and a pilot's reactions are automatic. I rammed open the throttles—a bit too fast. As we leapt into the air again the Sapphires stuttered and stalled with a tremendous rumble, and I was told that great flames shot from them. I throttled back a bit, unstalled the compressors, and the day was saved.

I allowed the brakes to cool and came in for a second landing. This was achieved, in part, by pulling off the runway where soft ground and grass helped to brake the 'plane. The ground crew were white-faced—as well they might be, for they had set the brake pressure at 120 pounds per square inch (suitable for ordinary, lighter Meteors) instead of at 150 pounds. I was not altogether blameless, for I overlooked the fact that I should have had a higher pressure registering. Further modifications and different engines turned this 'plane into the "jet lift" Meteor flown by the Westland Company in 1954.

As I have recorded, the Mark VII which I flew at displays and overseas went like a bird. By some crazy ruling in the contract, however, our figures were not accepted, and another Mark VII had to undergo an identical repeat of the plane's initial flight trials before it could be delivered to the Ministry of Supply for official clearance prior to going into production for the Services.

Many other contracts are equally lunatic. There is, for instance, the sort that causes aircraft to be built in quantity even after they have been proved useless. The contract calls for a specific number of 'planes, so that number are built—only to be tested, towed away and broken up. The cost? It falls on the taxpayer.

George Carter's original Meteor was designed in 1941. Today modified versions are still doing a useful job, and will probably remain in service for some years to come. The first Whittle engines installed in the Meteor were each of only 1,200-pound thrust, and no one realized how quickly they would be developed, and the number of guises in which the Meteor would fly.

Sixteen hundred-pound thrust engines soon came along, and these powered most of the Meteor I's which went into service with 616 Squadron to fight the flying bombs towards the end of the war.

Then came the Mark III, with 2,000-pound thrust Derwent I engines, which equipped several post-war R.A.F. squadrons. They were very manœuvrable, but at speed in rough air they "snaked" badly—the nose swinging from side to side. The controls were terribly heavy, and at cold temperatures over 20,000 feet the engines tended to "surge". To the pilot this felt like a rapid series of explosions which violently shook the aeroplane and, if allowed to persist, would wreck the engines through overheating and vibration. The cause was a breakdown of airflow inside the compressors at high engine r.p.m. and low air speed, and the simplest cure was to reduce engine speed. This resulted in reduced power and ability to climb, and as the 'plane went higher into colder air, the surge recommenced and the cycle was repeated. Often a Mark III's ceiling was reduced to 26,000 feet by this phenomenon.

Even today jets are not free from surge on cold days at great heights. Gun firing near engine intakes, or any oblique air flow such as occurs in a turn, may also cause surge or compression stall. Some engines are less prone than others, Metropolitan-Vickers' compressor being particularly good in this respect.

Some 220 of these early Meteors were produced before I joined the firm, but there was still work to be done on the Mark III. The Ministry of Supply were not satisfied with its spinning trials, and they had to be re-done. One of their test pilots had lost his life on these trials, and it was curiously morbid to do a job that had killed another, very competent, pilot.

I found the Mark III a cumbersome 'plane to spin. She did not want to spin, but preferred merely to drop a wing and her nose. By using full rudder she could be forced on to her back, where she would rotate slowly before she put her nose down and went into sickenly unsteady and drunken undulations. The stick tended to thrash about the cockpit and took a great deal of restraining. The 'plane shook and banged in protest, and the angle of spin was so steep that I was often convinced that her nose was beyond the vertical.

When recovery action was taken—full rudder in the opposite direction to the turning of the spin and stick forward—the nose seemed to drop even further into the dive and the speed of the rotation increased. Then, without warning, the 'plane would stop dead like a frozen top. My head was snapped to the side of the cockpit, and in a horrible, stalled silence I could look straight down to earth. If the halt came when the 'plane was on its back and beyond the vertical, I would be thrown forward against my harness.

In this stalled condition things were tricky, for despite having my head treated like a shuttlecock, I had to centralize the rudder quickly—or the 'plane would start to spin in the opposite direction.

I could never persuade anyone from the Supply or Air Ministries to occupy the rear seat of a Mark VII two-seater during its spinning tests, and I frequently invited members of Gloster's design staff, just for the fun of hearing their excuses. Generally they did not conceal that they considered themselves too important to face such risks.

I generally started spinning at 20,000 feet, finishing at 10,000. Further trials started at 40,000. The trials called for two complete turns for a fighter, four for a trainer. We usually went well beyond this, and on one occasion one of our pilots achieved, I think, sixteen turns.

The 'planes were fitted with a tail parachute for their first spinning trials. In emergency, if other methods failed, the 'chute could be streamed to assist in getting the nose down and checking rotation. I never used it myself, although I had cause to complain of the position of the parachute's release switch. To begin with, on the Mark III it was near the floor on the right side of the cockpit, and I argued that it was ridiculous to have to take one's right hand from the vital stick in a time of emergency—a last resort before baling out—in order to grope for the switch. Gloster's passed the buck. "Boscombe approve of it where it is."

"Boscombe have just lost a bloke," I replied, "and I want the switch on the left combing just above my throttle. There it can be seen, and reached by a simple six-inch movement of my not so important throttle hand."

Gloster's were adamant—the tail 'chute installations had taken months, and this would delay things still further. It was the old business: push, push, push, at the pilot's expense.

I replied: "All right, you can damned well get someone else to do it. I'm not, unless you put the switch where I want it."

To which they came back: "We've already had people fly with the switch in that position, so why shouldn't you?"

In the end, with ill grace, they gave way to me, and later the alteration became an official Boscombe requirement.

At the Central Fighter Establishment a flying man ran the show. He said what he wanted, and got it. But in the industry the men who fly the 'planes seemingly came second to schedules, and the chief test pilot, at the end of the line, was a necessary nuisance who must not be allowed to interfere in the practicalities of commercial enterprise.

The Gloster Gormless came and went, and plans were made to build a Meteor VI. Nothing came of the idea because, to some extent, my opposition was heeded for once. For while the Mark VI would have been an improvement on the Mark IV, it did not promise to be a sufficiently better 'plane to make its production worth while. Something basically superior and more advanced *was* required. (However, another 'plane on which I was keen did not go into production—either-a Meteor protégé of mine with sweptback wings and tail, and inboard-mounted engines.)

Not without its troubles came the Meteor VIII. This was in 1948. To start the transformation of the Mark IV into the Mark VIII a nose lengthened by thirty inches was added. An extra 95-gallon fuel tank between the gun bay and the existing tanks filled the space and replaced the 1,000 lbs.-odd of lead ballast it was necessary to carry in the nose of the IV. This raised the fuel capacity from 325 to 420 gallons, and shifted the 'plane's centre of gravity and altered the stability, for as the fuel goes its weight goes, and the centre of gravity moves forward.

The Meteor's ammunition, weighing some 800 pounds, was a disposable load, and was gone in about twenty seconds of steady firing. It, too, was moved thirty inches further forward. This variation in weight also seriously affected the centre of gravity: as the guns were fired the aircraft's centre of gravity moved aft—the dangerous direction. Consequently, the worst condition of stability occurred just after take-off with tanks nearly full and ammunition tanks empty. The danger became acute in turns or coming out of dives, when the unstable 'plane would "tighten" into the manoeuvre. A third worry was the fact that all 'planes tend to have stability and damping trouble with increased height.

The design office had worked out that the high-mounted tail of the second Gormless prototype could be used to good purpose on the Meteor. By a coincidental freak, it had the correct area, and gave the requisite lift and control to manoeuvre and stabilize the new, long-nosed 'plane.

The new nose and tail ideally complemented each other. Logically, therefore, the 'plane would have to have both the new nose and tail. With the two additions the Meteor VIII was a reasonable proposition as a fighter: with but the new nose only it was useless.

Understandably, then, I was flabbergasted to learn that the first hundred Mark VIII's would be fitted without the new tails. Apparently orders for ordinary Mark IV tails had been placed with sub-contractors, and they had to be used up before the new unit could be made available. So one hundred impossible aircraft were to go to fighter squadrons.

I hit the ceiling. "Turn out another hundred ordinary Mark IV's if necessary," I urged, "but not a fighter that's useless."

I was told, in effect, to mind my own business, and once again the buck was passed. "Boscombe will decide." But Boscombe's pilots agreed, as they usually did, so this time the Ministry of Supply passed the buck. "We'll send it to Central Fighter Establishment. They represent Fighter Command, and if they say they can use the 'plane that's good enough. We'll build them with the old tails."

I still had good contacts at C.F.E., and dropped a word to them. The IV-tailed Meteor VIII's were not built, but the story shows how the official mind works, for the M.O.S. were fully prepared to overrule their own testing establishment and myself in the matter—and pass on to the R.A.F. a most unsatisfactory aircraft.

The prototype Meteor VIII was smooth and pleasant to handle. It was faster than the Mark IV despite its extra weight: 15,200 pounds against 14,700. During trials I reckoned it was doing 605 m.p.h. at 5,000 feet, and I put this down to the cleaning-up of airflow over the improved tail. While the Mark IV pitched and bucked when approaching the speed of sound, the very first VIII simply showed slight trim changes and dropped a wing. It was lighter to manoeuvre and an easier spinner. Its ailerons were still heavy, however, and remained so until the new spring-tab type were introduced. In production, however, VIII's did not repeat the prototype's pleasant characteristics. This is a common fault—you test a hand-built prototype, clear it, then the mass-produced article isn't the same.

I flew night after night between October and December until the VIII was ready for her Boscombe Down trials. These were hardly under way before the Officer Commanding, Flying, took off in bad weather, crashed her in a snowstorm and was killed.

Thousands of VIII's were to be built: good 'planes—but still outdated by their American and Russian contemporaries.

The Meteor VIII was followed by the Mark IX: a similar 'plane, save that it was intended for fighter reconnaissance work and was fitted with cameras looking through glass windows in the nose. Then came the Meteor X, also a reconnaissance 'plane, unarmed, and designed to operate at high altitudes. Because of this it carried the old, long, rounded wings and its speed down low was limited to 500 m.p.h. Arguing that it did not need to manoeuvre like a fighter, the authorities kept the Mark IV tail—despite the longer nose with its extra fuel.

It was heavy and uninspiring to fly, and saw daylight when I was in the Argentine. The first one folded up, killing my head production pilot. I never discovered why he was in it when an experimental pilot was available, nor was the cause of the disaster fully established, although it was believed that a camera cover over a porthole in the fuselage may have come partly adrift and broken the glass. This would have served to scoop air into the fuselage and blow it up like a balloon.

We produced a few Mark VII's with the new tail for special duties. We dubbed them "Seven-and-a-halfs". They were lovely machines, but the M.O.S., for reasons best known to themselves, did not agree to a general conversion being made.

As chief test pilot, I did all first or tricky flights. I felt that since mine was the responsibility, mine should be the neck risked. There could be no excuse if a wrong conclusion, due to a pilot having only a fraction of my experience, led to a fatality. Consequently I felt in no way responsible for the three unhappy fatalities that occurred during my term of office. One was a combination of engine failure and circumstances; the second the breaking-up of an aeroplane; the third an unfortunate aerodynamic fault.

Squadron-Leader W A. Waterton, A.F.C., at the time of the R.A.F. High Speed Flight, Tangmere, Sussex, 1946. (Drawing by Cuthbert Orde.)

The Gloster E1/44 "Gormless" (*above*) taxiing out for her first flight, and (*below*) in flight east of the Malvern Hills.

A captured Junkers Ju 388 with Focke Wulf FW 190 in the background. Typical of enemy aircraft tested and used in trials against British planes by the Air Fighting Development Unit.

The rema of the rocket phoon afte tyre burst take-off, V tering, Lin 1944.

Tempest V JN 757, damaged during the author's first flight in this type. An obscure fuel system defect caused engine failure which led to a forced landing.

The author in EE 528, with "Gertie" and "Ermintrude" in the background.

(Below) Pilots of the R.A.F. High Speed Flight, Tangmere, 1946. Flight-Lieutenant Neville Duke, Group Captain "Teddy" Donaldson (the C.O.), and the author.

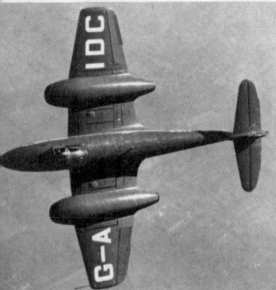

(*Above*) The last of the Glad
tors flying—Brockworth, 195?

(*Left*) AIDC—the P V carmi
and white Meteor IV under te
prior to its tour of Scandinavi
This aircraft was damag
severely in Belgium, and par
of her went into the fir
Meteor VIII.

Istanbul, spring, 1948. President Inonu
(*left*) with H.M. Ambassador, Sir David
Kelly, and the author.

(*Above*) Ciampigno Aerodrome, Rome, in spring, 1948. (*Below*) After the abortive 100-kilometre world record attempt which nearly ended in disaster.

Athens, spring, 1948. King
Paul, ex-King Michael of
Rumania, Queen Frederika of Greece, and the
author.

(Below) Prince Bernhard
of the Netherlands has
his first flight in a jet
from Pershore aerodrome.

GLOSTER METEOR MK VII ROLLS ROYCE DERWEN

Tom Sopwith
tries out the first
jet trainer.

(Below) AKPK—the
first Meteor VII; the
first two-seater jet
trainer in the world.

(*Above*) Hare and Tortoise. The Meteor turns its letter over to the heli-copter at Orly Aerodrome, Paris, 27 minutes out of Biggin Hill. (*Below*) Paris, spring, 1948. In front of the nosewheel door, General Léchères and the late General de Lattre de Tassigny; General Hartman talking to the author. These, and other notables, all flew in G-AKPK.

(*Top*) Why aeroplanes don't behave in flight as models do in tunnels. Note the skin deformation on this Meteor at just over 500 m.p.h. Designers can pooh-pooh this, but at least they can be shown that it does happen.

(*Centre*) WA 820, the Sapphire-engined Meteor which nearly came to grief when the brakes failed to stop her.

(*Below*) The first Meteor VIII flying in company with a high-tailed prototype of the "Gormless." The high tail from the latter was married to the Meteor fuselage.

(*Left*) After the displa
the CF-100 in Otta
March, 1950. Edgar A
Chief Engineer of A
Canada; H.E. the G
ernor-General, Visco
Alexander of Tunis;
Hon. Brook Clax
Minister of National
fence; the author; Pr
Bernhardt of the Net
lands; H.E. the Net
lands Ambassador;
A. V M. "Art" Ja
R C.A.F.

(*Right*) Boston air
display 1950. Cap-
tain "Chuck" Yea-
ger the first man
to fly supersonic-
ally and the author.

(*Left*) The start of the flight
which "the lid" came off. Av
Canada's chief test pilot, D
Rogers, is in the rear cockpi

(*Top*) CF-100 in flight.

(*Centre*) The first Javelin prototype in flight.

(*Below*) First take-off of Canada's first modern home-designed and built jet fighter, January 19, 1950.

9

BELGIUM, HOLLAND AND THE ARGENTINE

Despite post-war American influence in many of our traditional mar-kets, hundreds of Meteors were sold overseas. The Argentine was our first customer, France—with, initially, one Mark IV for experimental purposes—the second. Then came Belgium and Holland.

I took the first Meteor to Belgium in October, '48, with their fighter ace, Colonel Donnet, in the rear seat. It was a trainer, and the Belgians (who, incidentally, produce really first-class pilots) were delighted with it. After being handsomely wined and dined in Brussels, I returned to England by B.E.A. Viking—and was given one of the worst scares I have ever had in an aeroplane.

There were few passengers, but one, a Scot, was "fleein' drunk" in both senses of the word. The Belgians were reluctant to allow him through Customs, and aboard the Viking his conviviality increased at the expense of the stewardess's bar.

It was fine all the way to the English coast, but at Dagenham we ran into black, cumulo-nimbus clouds with their attendant rain, electric storms, lightning and rough air. The approach into Northolt was by instruments.

It was a nightmare ride. At one moment the engines roared full out—then, a few seconds later, came dead silence as the throttles were yanked back. My jaw would fall on my chest under G—then, in a flash, the reverse occurred, and only the safety belt prevented me from being thrown up and hitting the ceiling. We circled, the din of the engines telling me we had missed the airport. Another terrifying try. This time we broke cloud in light rain at well under 800 feet to find ourselves skimming the red warning lights atop the hangars on the north side of Northolt. We shot upwards again, and the Viking hit a colossal bump. I am certain it was the slipstream of another aircraft.

By this time the Scot was as cold sober as yesterday's porridge. The face of the stewardess was grey; doubtless mine was, too. I said to her: "Your bloke up front has a bad case of twitch, hasn't he?" To add to my perturbation, she replied: "I'm terrified. He's already got one prang to his credit. . . . I'll be glad when we're down."

We made it on the third attempt. After an appallingly jerky approach we landed half-way down the runway and ground to a halt, lucky to stop before over-running it. I was never so glad to climb from an airliner, and if I'm any judge, the pilot felt the same way. I learned later that he and B.E.A. soon parted company.

As a pilot it is difficult to sit in the back and suffer silently someone else's bad driving. By the time a 'plane is airborne I reckon to be able to form a fair opinion of the man flying it. The way he taxies, brakes, turns, and uses his engines are usually sufficient to tell me whether I'm in the hands of a smooth or rough pilot. (In the same way, I claim, when a car passenger, to tell what sort of a pilot the driver would make.) Looking back, that was my worst civil airline flight. B.O.A.C., on occasions, have had me apprehensive, and American airliners have caused me worry. On the other hand, I have no complaints of my trips by French, Belgian, Dutch, Norwegian and Canadian lines. On five North Atlantic, and five crossings of the North American continent, T.C.A. has afforded me not a moment's concern—although their passenger handling is noticeably rougher and more offhand than the British.

I spent a lot of time in Belgium in 1948/9, for it was my responsibility to get their fifty Meteors across. Two lots of two were flown over in December, '48, and by April '49, leaving one pilot at home to look after what little production had not been dealt with, four of us were making two deliveries each a week. The deliveries followed a fairly fixed pattern: take the aircraft across in the late afternoon, enjoy the pleasures of Belgium, then return next morning (usually by the firm's Rapide or Sabena) with a few purchases unobtainable in England.

The Customs at Lympne treated us decently on the whole. One day, however, I found a bright young Customs officer at the Rapide with a screwdriver. "What the dickens are you up to?" I asked.

"I'm entitled to inspect the aircraft as I see fit," he retorted.

"Quite," I said, "but have you a ground engineer's licence?"

A puzzled: "No."

"Right: if you touch that 'plane it becomes unfit to fly, and we're on your hands until a qualified man puts it right."

His youthful ardour cooled.

One trip to Belgium provided an amusing incident. My number one pilot, Rodney Dryland, and I climbed aboard the Sabena Dakota bringing us home, with our parachutes slung over our shoulders. The stewardess took a dim view; the Captain's was even dimmer. We could see the point—it was psychologically bad for the other passengers. The Captain brought a high-level official along. By this time we were in our seats, the 'chutes in the racks above, grinning mischievously at the other passengers, who were eyeing us nervously. Following the airport official, who was in a foul temper, were the Belgian crew. Then came the police, who insisted that we could not stay aboard with our parachutes. Either they were stowed in the baggage hold or we did not travel on the 'plane. We haggled and argued good-humouredly for a while, then gave in. The 'chutes were removed. I'd rather have had one on the B.E.A. trip.

Overlapping our deliveries to Belgium were our dealings with the Dutch. Prince Bernhardt's visit led eventually to an order for thirty Meteors.

The first to go over was the veteran Mark VII in which I had notched-up a hundred flying hours, had taken to Turkey, and to which I had grown very attached. I never forgave the Dutch for smashing her up soon after I left: she was landed short and a ditch tore off her undercarriage.

My return to England in a Dutch Air Force Dakota was delayed by violent gales. Buildings and hangars were badly damaged, and the station commander was in a dreadful tizzy—not because of the threat to the aircraft and the discomfort of his personnel, but because of the number of guilders going down the drain. This, alas, I found to be an unfortunate Dutch characteristic: they reduce everything to currency.

Eventually it was arranged that Belgium and Holland would build their own Meteors: the airframes at Fokkers' works in Holland, the engines (under Rolls-Royce licence) at the Belgian national armament works. The two countries did not appear to have much time for each other, and I was constantly asked by the Belgians to fly the Fokker Meteors to confirm their views that they were inferior to Gloster-built 'planes. I avoided that one like the plague, and told my pilots that under no circumstances were they to fly them. Otherwise there would certainly have been a monumental three-cornered row.

When, in 1947, the Argentine bought a hundred Meteor IV's, and sent a dozen of their pilots to England to learn how to fly them, their training came under my ægis. It was a task that brought back many memories, for I had left an R.A.F. fighter squadron to be a flying instructor, and in two years more than six hundred pupils had passed safely through my hands. More than nostalgia, it was an exciting challenge, for I did not agree with many official training methods, and this provided an opportunity for employing my own ideas. It was not easy, however. We were in no way equipped as a training school; certain of my staff were unco-operative; and—greatest difficulty of all—the two-seater trainer was still a long way away.

I had to have pilots' notes translated into Spanish, since only one Argentine spoke reasonable English. From Second Lieutenant to Captain, their ages ranged from twenty-one to thirty-six. They had from two to six hundred flying hours, but had never flown at more than 220 m.p.h. and were completely new to modern aeroplanes. And I had only six weeks in which to prepare them for a formation display over Buenos Aires.

They started at Rolls-Royce, where they learned the technicalities of the Derwent engine. Then they went to Air Service Training to fly Ansons and Oxfords and so adjust themselves to twin-engined aircraft. To the disbelief of the A.S.T. staff, I insisted that the Argentinos should fly the 'planes in the manner of Meteors, taxiing on brakes rather than engines, operating engines against brakes to begin the take-off, using partial flap and engine to simulate the Meteor's relatively flat approach, and landing tail-up, as with tricycle undercarriaged aeroplanes. These, and other Meteor features, I myself taught instructors and pupils, for they were not interested in Ansons and Oxfords. I have always maintained that one aeroplane can be flown to show the characteristics of another, quite different type.

At Gloster's they learned to run the jet engines on the ground, and discovered the intricacies of undercarriages, flaps and air brakes—innovations lacking in their own, pre-war fighters. To instruct them in taxiing, we removed the canopies from the cockpits of Mark III's and sat cowboy fashion on the ammunition boxes behind our pupils, yelling instructions above the roar of the engines, and signalling by hand.

As the taxiing, braking and engine handling improved, we aligned the aircraft down the runway and raced them faster and faster until the controls began to take effect and the wings bore most of the aircraft's weight. We made them lift the nose-wheel near take-off speed, gently lower it again, then brake to a halt. They had virtually done a take-off and landing, and it was now up to them in the air. We could do no more.

My good right arm during the training was Rodney Dryland. Sitting atop a Meteor, red hair streaming in the wind, he was a wonderful sight. He was the pilot killed at Moreton Valence when I was in the Argentine, and I felt his loss keenly, both as a pilot and a friend. No other Gloster flyer was quite like him.

At last the time came for the Argentinos to fly the Meteors. There was tremendous excitement as they took the air. If we could not be with them physically, they carried with them our prayers and hopes.

In his pocket, each pilot carried a note I had drafted. It said: "To Whom It May Concern—This will identify X, an officer of the Argentine Air Force. He speaks Spanish and will understand Italian. Please accord him every possible courtesy and assistance. If found lost, strayed, or otherwise in difficulty, please ring Gloucester 66294—Chief Test Pilot."

Although there were incidents and emergencies—an engine failure, overheated brakes, fuel shortages in the air—the telephone rang only once. We collected the undamaged 'plane from another aerodrome, the pilot telling us that his compass had gone wrong.

Before they left, the Argentinos flew in a formation that would have done credit to the R.A.F. I felt proud of them—and of myself. There is quite a combination when a fighter pilot blends Latin temperament with Argentine pride.

England sold the Argentine more than Meteors. Their pilots bought M.G. cars and B.S.A. motorcycles from their generous allowances. One day, en route to the aerodrome, I heard a thunderous roar of engines. Headed by an M.G., a formation of bikes tore past me at high speed— keeping to the right-hand side of the road. Like a squadron of fighters, they changed from a V-formation to line abreast, then to line astern, regardless of the goggle-eyed motorists towards whom they raced head-on.

The police sorted things out amiably enough.

The Meteors were shipped to the Argentine. Then, as the months passed, disturbing reports arrived from our people over there: testing the reassembled 'planes was going far too slowly, the job was costing far more than it should, and the Argentine authorities were unhappy.

One morning in May, '49, the general manager called for me. "I'd like you to go to the Argentine," he said, "and clean up the mess there. Finish flying the Meteors and sweeten up the Argentinos. The service manager tells me that all the 'planes will be ready to fly by the beginning of next month."

I thought for a moment. "I'd better allow six weeks," I said. "A fortnight to fly and get the Meteors off test, a fortnight to visit round the Argentine Air Force and sweeten up the authorities, and a further fortnight to cover any setbacks and complications." I had heard of South America's *mañana*.

I forgot, however, that in the song "*mañana* never came". I was to spend three months in the Argentine.

I drew up detailed instructions to cover my absence. Rodney Dryland would be in charge of the overall administration of the flight test section: flying control, the aerodrome, pilots, discipline and the detailed running of production and overhaul testing. I was confident that Dryland's drive would keep them moving, and that if things went wrong it would not be from want of trying.

I flew to Buenos Aires in a York belonging to British South American Airways, now incorporated into B.O.A.C. It was comforting to see that the crew had war ribbons on their chests. Despite the purists' arguments that R.A.F. pilots are less suited to airline flying than their own, specially trained, people, I'd put my money on the ex-R.A.F. man every time: the flyer who has been through the Air Force mill, has ducked flak and fighters in all conditions and weathers, has brought back shot-up and crippled aeroplanes—and has fought through civil airline prejudice to bludgeon his way into their jealously guarded Freemasonry.

On my arrival in the Argentine, I soon found that it was out of the question to make a direct approach to the Argentinos. A peculiar mixture of pride and face-saving made it impossible for them to be told anything. One had to make a slow, circuitous and circumlocutory approach to a problem: a tedious, wearying and exasperating method of doing business.

The rôle of our agents there was really a political one. They knew little of Meteors, but plenty about the Argentine—which doubtless justified the reputed £400,000 commission they drew on the sale of the hundred Meteors. Gloster's head representative was Vic Hewlett. He was twenty-five years old and directly responsible for reassembling the aircraft after their delivery, and for seeing that the Argentinos received the servicing help they required. He had learned his job properly, having joined Gloster's as an apprentice, and was extremely keen, hard-working and honest. Frankly, I do not think the firm appreciated his true value.

Hewlett had a tough job: doing business with an awkward and suspicious customer, coping with such problems as a member of his staff kicking an Argentine sentry, a firm eight thousand miles away which did not fully appreciate his difficulties, and a surly crew of workmen who considered themselves underpaid and generally hard done by. A lesser man would have thrown in his hand.

The Meteors had been built, tested, inspected and broken down at Gloster's, then crated and shipped to the Argentine with spares. It was when they were unloaded that the trouble started. The aircraft were subject to Customs duty, despite being for the Argentine Air Force. Funds had to be diverted to pay the duty, and things were at a standstill until the money came along. The shipment was not considered at all urgent, and was just left lying in the Customs shed. Spares totalled twenty per cent. of the entire shipment, and this figure was normal. More would have been justified by the amount of damage suffered by the aircraft *en route*: damage when the aircraft were dismantled and packed, damage caused by badly battered crates and bad packing, damage resulting from careless handling by dockers, and damage due to the corrosive action of sea water on the journey across. This damage, due to British workmen, at a time when we were desperately trying to boost exports, totalled scores of thousands of pounds.

The Argentinos were not prepared to pay for the Meteors until they had been tested and flown, and when we complained to them that we could not prepare the 'planes for flight until they were released from Customs, they brightly replied: "We often have stuff in Customs for seven months or more."

The contract said that the Argentine was to supply us with help in the form of manpower, oxygen, paraffin, oil and battery-charging facilities. But the contract did not specify *when*, leaving it to an Argentine goodwill that was sadly lacking. When we wanted battery chargers, they were already working to capacity; when we needed fuel the bowser was away fuelling Argentine 'planes—and so on. Small wonder, perhaps, when Argentine mechanics had been chased away as "clots" by a former representative, and the Argentinos found their Meteors carried retreaded tyres and worn cannon. (These were to Ministry of Supply specification, as supplied to the R.A.F., and Gloster's were not to blame.)

When I arrived at the airport of Moron I was appalled at what I saw: a graveyard of heartbreaking neglect. Several wingless Meteors stood parked in the open in an obscure corner. There were no covers over them. Controls, engines, pipes and tubes lay exposed to the elements. The heavy steel bolts and forgings which secured the outer wings were bronzed with rust. Insects had built colonies in the vital pipelines. Most comic—had it not been so abominably tragic —was the condition of the tails. The Argentine swallow—the *hornero*—is an adobe builder, like the barn swallow in England, and back to back, at the juncture of each tail and fin, eight birds had built nests. The outer wings of the 'planes lay corroding on the ground. At best the Meteor's ailerons justified criticism, and there they were flapping loosely in the wind, together with their trim tabs.

I went off the deep end and raised the devil. Hewlett agreed that the situation was infuriating, but assured me that he simply had not been able to get the Argentinos to do anything: all his protests and cajolings were airily waved away. I tried myself, with no greater success. The officer in charge of the acceptance and inspection for his Government expressed his regrets, but feared that he saw no way in which to assist me. He had neither the manpower nor the storage space.

Appointed to test the Meteors when their reassembly was complete was young Lieutenant Dominguez. He was keen to learn, and not above asking or taking advice. Unfortunately, however, he saw little wrong in the way the aircraft were parked.

"What happens to these aeroplanes when the O.C. Overhaul and his staff have finished with them?" I asked.

"I will fly them, make adjustments, and send them to the squadrons."

"My poor chap," I said with feeling, "I wouldn't touch your job for a million pesetos."

He jerked upright. "Why not?"

"Well," I replied, "that rust on the wing roots, for instance. What will you do about it?"

"We'll emery-cloth it off."

"Good God!" I cried. "Those are specially fitted bolts. If you do that the clearance will increase, the bolts will be loose and the wings will flap. And with the hammering the ailerons are getting," I added, "you'll never get a 'plane in lateral trim. Don't forget, too, that with all those bugs and things you probably won't have any instruments working or trustworthy." While he was reeling under this, I added: "My dear chap, you may have a future, but it won't be a long one—if you fly those."

Three days later the 'planes were moved under cover and cleaning-up began.

That set the pattern for subsequent achievements: if we wanted anything done we had to put someone on the spot. Whether it was oxygen, fuel, or anything else, we could get it only by intimidation and by refusing, in turn, technical assistance, spares or special tools.

Considering the precision work that goes into the manufacture of aircraft, and the subtleties of jet flight, it is remarkable what can be got away with in 'planes. They continue to fly despite lack of servicing and attention—even when badly designed or assembled in the first place. But there is a limit, and then the inevitable occurs: a prang. There were many in the Argentine, and although they upset me, I was hardly surprised.

A mixed-up people, the Argentinos. A fusion of Italian, Spanish, German, French and British bloods have produced manual skill, dexterity and first-class brains confused by corruption, and lack of discipline and direction. I have seen an Argentine private doing half a day's gruelling punishment drill under the blazing sun for failing to salute an officer a hundred yards away—while soldiers with rusty weapons, a two-days' beard and soleless shoes raised no disciplinary issues.

Among my visits was one to the National Technical Institute—the State aircraft factory—near Cordoba. It was a haven for ex-enemies of Britain. Former allies, too. Aircraft were designed and manufactured there. It was a remarkable set-up. Everything was produced on the spot—instruments, propellers, airframes, engines, Perspex, plywood: all came from the 6,000 workers in the Institute. Most impressive and ingenious—but it did not bear too close an inspection.

Under Argentine direction, the technical staff were mostly Germans, Poles and Italians. Many of Focke-Wulf's staff were there, including designer Dr. Rueth, and Willi and Kurt Tank of 190 fighter fame. Ulrich Rudel, the renowned one-legged Stuka pilot was at the factory: I sat in on his interrogation at Tangmere in 1945. There were many Poles, some of whom had worked at Farnborough and Boscombe Down, and I also met Major-General Adolph Galland, Hitler's top fighter ace, soon to be head of the new German Air Force.

10

THE MIDDLE EAST AND FARNBOROUGH

I was not long back from the Argentine before I was travelling again. During my absence Egypt concluded a deal for fourteen Meteors: two Mark VII trainers and twelve Mark IV's (the first of an eventual order for thirty-four aircraft), and it was my responsibility to get them out there. I would select and fly with the initial delivery, a VII and a IV, prove the route, and prepare the way for those who followed. I elected to take the trainer, with Gloster's veteran mechanic, Freddy Cook (who would stay out there as our servicing representative) in the rear cockpit. The pilot I was putting in charge of subsequent ferrying to Egypt flew the Mark IV. I prepared my detailed route, and we set out on a sunny morning in October, 1949.

Our first stop was at Nice, an hour and twenty minutes out of Farnborough, where we were greeted by Jack Lycett from our French agent's office. Half-English, half-French, he had been a wartime British paratrooper and was quite a lad. Ours were the first jets Nice had seen, and this, coupled with their Egyptian markings, made them "news". We were soon buttonholed by an intense young lady who wanted us to broadcast over the local radio.

With Egypt licking her wounds after the war with Israel, and the Near-Eastern atmosphere still electric, I was reluctant to advertise that we were selling the Egyptians Meteors. Yet France was still a prospective customer, and Lycett felt that a bit of free publicity would do no harm.

So I agreed, first buying the girl a drink and asking for time to eat. Throughout the meal she kept urging us to hurry to the local church where a colleague was standing by with his microphone, and although we speeded up, a choir was occupying the air when we arrived. The lady apologized, and we repaired to the vestry to fortify ourselves with some excellent wine we found there. Later the clerics appeared, and gave us dirty looks. Their lack of Christian charity struck us as being a pretty poor show.

The choir ended their songs of praise, and our intense young lady's great moment arrived. After formal introductions she was to conduct a question and answer interview—but it didn't work out that way. Lycett quickly seized the initiative and, anticipating the questions, gave his listening audience a potted lecture extolling the Meteor's virtues. Although she was a pest, the girl really didn't deserve it.

The following day, on the way to Malta, we ran into trouble. The Mark IV's engines started to "surge"—that dangerous rise and fall in revs, temperature and power which, if not quickly attended to, can soon ruin a 'plane's engines. My other pilot's aircraft, nearly a thousand pounds heavier than our two-seater, was feeling the effects of height more than we were. Although I was well throttled back, he could not keep up with us. I slowed down to stay with him.

Troubles, it has been said, never come singly. They didn't then. Calls from my number two told us that his wing and belly tanks were not feeding the main ones. On mains alone Malta was impossible, and although I could hear Rome and Naples over the radio, I could not raise them. Then the other pilot's radio started to go: fortunately only on transmission. By signals we established that he could hear us, and I decided to put him down at Palermo, in Sicily, then carry on to Malta to collect fuel and assistance.

There was one alternative, however. We dropped height from 36,000 feet, and I told number two to depressurize his cockpit. The Meteor fighter's cockpit is pressurized with air tapped from the engines' compressors, and this supply of air also transfers the fuel from the external tanks to the mains. There was a chance that by depressurizing the cockpit, extra air would be available for the fuel system and would do the trick. It did. At about 14,000 feet he signalled that his fuel was transferring again. We climbed back to 23,000 and made Malta in the middle of violent thunderstorms.

As I went towards the watch-tower to report I bumped into an old acquaintance—but the last person I wanted to see at that moment. He was Flight Lieutenant Jacobs, an excellent chap who had been Observer/Navigator to night-fighter ace Wing Commander Bob Braham. He was, however, a Jew, and there we were, just after the "troubles", with two aircraft bearing Egypt's Islamic insignia! But "Jake" merely grinned and saw the humour of it. He promised not to blow us up before we left, nor to cause our 'planes to disintegrate in the air—although Jacobs, in charge of Air-Sea Rescue, said he had a Lancaster standing by to drop us a lifeboat. We've often laughed since over the encounter.

After the R.A.F. serviced and repaired the Mark IV, we flew on to El Adem via Tobruk. I was amazed at the number of ex-Africa Corps men working there. God knows what they must have been wanted for at home to have chosen voluntary exile in that forsaken hole.

We arrived at Cairo soon after dusk on the third day after leaving England. There was plenty of fuel left in our tanks, so we introduced Egypt to her first jets by beating-up the city in close formation at low altitude. When we landed we were greeted by an imposing and happy array of military and civilian brass. As delighted as any was our agent: understandably, since our two aircraft represented to him between three and four thousand pounds in commission.

On the following evening we displayed the Meteors to the Prime Minister, the Minister of Defence and their retinues. The show followed the familiar pattern. Even now the Meteor's high-speed aerobatics are impressive: in those days they were phenomenal. When I came down, the Prime Minister congratulated me and said: "It was wonderful. I want to see Egyptian pilots doing it just as well in a week's time."

I replied that I didn't see why they should not be flying the jets in a week, although I doubted the wisdom of their doing beat-ups with such limited experience. The looks and gasps which greeted this made it clear that "Yes, Excellency," was the only reply ever to be given to the Prime Minister of Egypt.

The Egyptians took over the 'planes for acceptance tests and a check of equipment and stores, and while these were in progress we made the usual tourist visits. Our guide kept us highly amused by assuring us in war-acquired English that "You'll get good gen from me O.K. I'll give you the griff on all this stuff. No duff gen."

After three days the other pilot returned to the U.K. while I stayed on to instruct the Egyptians. My work was by no means easy, for my Egyptian pupil pilots not only knew little of jets, but were short of experience with twin-engined aircraft. Their fighter squadrons were mainly equipped with single-seater Spitfires and Italian Machiis and Fiats, for with the disastrous Israeli war a vivid memory, the Egyptians had bought up any fighter on which they could lay hands, as did the Israelis.

Their pilots, however, were a decent, steady, helpful lot—although generally lacking the "push on" enthusiasm of the R.A.F. They were a mixed bunch, and a surprise to someone new to Egypt: blonds of Armenian extraction with Persian-sounding names; lean Semitics; solid roundheads of Turkish origin; and the "true" Egyptians—the fore-head-bulging Copts.

One or two were square pegs in round holes—possibly as a result of the Egyptian Air Force having been modelled in part on the R.A.F. There was, for instance, one pupil who had been a bomber pilot, and who was clearly not at home in Meteors. He had the honesty to tell me so, and I sent him back to his heavies, and I've since wondered how many pilots have gone to early graves for lack of such candour.

The Egyptian N.C.O.'s were excellent workmen, abominably underpaid, but abounding with skill in improvization. I saw this at Almaza, where the Americans had left a graveyard of about a hundred transports, mainly large, twin-engined Curtis Commandos. All were in a state of abject dilapidation, but the "Gippos" had rebuilt forty useable 'planes from the abandoned junk. Some, with other aircraft, were employed in most unlikely rôles.

With pride, the Egyptians showed me, among their Dakotas, two Emmet-like jobs dedicated to the massacre of Israelis. Among British stores from way back, the Egyptians had obtained light bomb carriers which had been fitted to pre-war biplanes. Each of these "toast-racks" could carry four small bombs, and two racks were usually fitted beneath a biplane's lower wing. On the Dakota, however, they stretched literally from wingtip to wingtip, and in places were double-banked. She must have carried hundreds of small eggs.

The other old Dakota was fitted as a ground *straffer*. Windows on each side of the fuselage had been removed, and extra ports added. Along each side would stand a row of sub-machine gunners, and the idea was to fly at low level alongside enemy troops, firing a tremendous broadside. I don't know if this infernal machine was ever used, but I dreaded to think of what would happen if over-excited gunners failed to cease fire when a wing got in the way of their sweeping spray of bullets.

A Spitfire, too, had been ingeniously altered. Of an early design, its single radiator originally hung under one wing. The Egyptians had re-designed this and it was now buried in the underside of the wing, with air ducted to it. The result was a Spitfire V much improved in appearance and I was assured, twenty miles an hour faster.

Since fuel and oxygen were short, I thought I would save time on the Mark VII and teach my pupils about twin-engined and tricycle undercarriages in one of the de Havilland Doves that stood about the tarmac. Unfortunately, I had never flown a Dove, neither flying notes nor instructions were available, and the de Havilland agent could provide me with nothing more than an illustrated sales brochure. The best I could do was crawl around the Dove with an N.C.O. and trace through the levers and switches with him to learn what each did.

I took her up to check quickly her flying characteristics. She was a disappointment. She lacked positive elevator control at slow speed. With the stick full back on take-off, nothing happened—absolutely nothing—until, at 80 to 85 m.p.h., control came with a bang, the nose reared up, and the 'plane positively jumped into the air if not checked. Elevator control during landing was far from positive, and lateral control as heavy as in a large airliner—and almost as responsive. She had twin engines to be sure, but on one her performance was pitiful. The Dove was comfortable, had a nice overall design and a good turn of speed but, like so many 'planes, was far from "complete". I hope it has been improved since, for there was much about it that I liked.

Anyway, the Dove served my purpose, and I was able to show the Egyptians how to handle twin engines, how to get in or go round "on one" if the other failed, and how to cope with a tricycle undercarriage. When they moved on to the Meteors, my pupils found their performance startling, and progress, although steady, was slow. It was hard for them to adjust themselves to the high rate of climb and speed, and the rapidity with which they could get lost or run out of fuel.

Since we used up to 300 gallons of fuel a sortie—2,400 gallons a day—I was not surprised when supplies ran short and we were temporarily grounded. I filled in time with more lectures.

Our agent left for England, leaving me his house, servants and car, but I could not take advantage of them. I fell sick with dysentry, and although it was November, temperatures were in the nineties, and I found myself too feverish to do anything but work and sleep.

I would have liked to have done a bit more for my pupils—who were flying creditably enough—but after three weeks I was ordered home. Before I left I attended a cocktail party at which high Egyptian government officials were present together with top officers of the Services, and I was afforded a first-hand glimpse of one of the reasons we were chucked out of the Canal Zone. A British couple were there: he a small, rotund ex-senior Army officer in his late fifties; she a vast woman whose size was in direct disproportion to her tact. In a voice that matched her bulk, she thundered at me: "Hear you've brought some Meteors out to these wogs!"

"I beg your pardon."

After a bosom-quivering breath, the remark was repeated—louder.

"No," I replied, "we've sold Meteors to the Egyptians."

"Oh! . . . How many,"

"A number."

"What sort of delivery are you giving?"

"Quite good, I believe."

"Huh! Typical Gloster answer. Just the same as when I was at the Ministry of Aircraft Production with Lord Beaverbrook during the war."

Her attitude may have helped to produce aeroplanes, but it didn't assist us in the Canal Zone. Yet I found her typical of many Britons living abroad.

The B.O.A.C. Argonaut bringing me home was to have left at seven a.m., but when the captain checked his engines at the end of the runway, one of the Merlins was pretty rough, while another kicked and spat, jolting the entire 'plane. We were told that owing to a slight fault in the "radio equipment" we would be delayed, and were given breakfast. Outside the restaurant the sight of the engine cowlings being removed and the irregular running of the Merlins suggested that the Argonaut's radio was oddly situated.

Halfway through the meal, plates and uneaten food were whipped from under our noses and we trooped back to the aircraft. This time the captain was taking no chances on magneto (or should I say radio ?) trouble, and did not run-up his engine until the actual take-off. It scared me out of my wits.

Our stewards seemed more interested in tips than service, and when an elderly gentleman became acutely distressed they could not have been more disinterested. An American woman passenger went to his assistance, and must have formed a pretty poor opinion of British airlines. Give me stewardesses every time.

During that winter—1949/50—a dozen Meteors were ferried to Egypt. Then politics reared its head, and in October, '51, an embargo was clamped down on arms to Egypt. The embargo was not lifted until August, '53, when we supplied Mark VIII's to complete the order. To expedite delivery of the first dozen of these VIII's, two Egyptian pilots came over to fly them back to Egypt in three trips with two of Gloster's pilots. At Rome on the first trip, one failed to take off: suspected loss of power on one engine. It was repaired and finished its journey, but in the meantime I was instructed to hold back the other eight. The reason, I gathered, was that holding up the fighters would strengthen Britain's hand in the Canal Zone evacuation negotiations.

"Tell them you haven't got the pilots, old boy," I was airily instructed.

"Fat lot of good that will do," I replied, "when their chaps have been to Moreton Valence and have seen how little we were doing. They know we've pilots to spare."

"Then tell them the 'planes aren't ready."

"When they've seen them lined up on the tarmac? Why not suggest that the Foreign Office does its own telling."

"Can't do that, or the Government won't grant us export licences."

The glorious pay-off was that the Meteors *did* eventually go to the Middle East, but instead of bearing the Crescent and Star of Islam, they carried the blue Star of David: insignia of the young Israeli Air

Force.

The Israelis who came to Gloster's—some of the top brass in their young Air Force—were an impressive bunch. It was not often that we received customers who asked about such matters as "amortization" and "utilizations". They came with the benefit of R.A.F., French, Czech, American and other training, and were young and dedicated. And some certainly relished a pork chop. When, at a luncheon, the sales manager and I carefully chose beef and chicken, they ordered pork, one of them explaining: "When in Rome . . ."

Among them was an ex-Farnborough technician and armaments expert who was said to have an excellent record of demolition and other blasts to his credit. Their C.-in-C. might have been a product of an English public school. He was slim, tall, blond, blue-eyed and slightly toothy—with an accent to match.

When it came to armament trials, and the Israeli stores came, the Gloster people nearly died of nervous prostration. They were not used to dealing with five-inch live rocket heads—and had no intention of doing so. Undaunted, the Israeli armament type arrived with his own spanners, and it took a great deal of time persuading him that in case of trouble our insurance people wouldn't look at us as a risk if the 'planes carried explosives. Finally, after much haggling, we used dummies with concrete heads.

Our pilots who delivered the Meteors were treated royally, although the Israeli Foreign Office kept an eagle eye on them, and they had little freedom of movement. Understandably, perhaps, since Israel was on a war footing. Elaborate precautions were taken from Cyprus onwards, special call signs and routes being used. The Israelis said that if there was another war with their neighbours they would not worry unduly as long as the odds against them were not more than five to one.

There was considerable worry over ferrying Arab aircraft out East. Some years before in Rome—a H.Q. of Jewish intrigue at the time—a grenade was found tied clumsily to the aileron of a Fury. It hardly seemed the work of Stern gang experts—but was perhaps a warning.

On another occasion some Meteors bearing Islamic markings were approaching Nice on their way to Damascus, when the R/T revealed frenzied activity among a number of Mustangs on the ground there. All seemed very anxious to leave, and as the Meteors went in the last of the Mustangs scooted out towards Italy. It was seen to be carrying the Star of David. At Rome the Meteors caught up with the Mustangs. Flown by Swedes, they were ex-Swedish, ex-American 'planes bought by the Israeli Air Force. The pilots had a get-together, and the Swedes revealed that they had thought the Meteors to be flown by Arabs—hence their haste at Nice.

The Gloster pilots found Syrian "tourist attractions" rather curious. One day there was a commotion near their Damascus hotel, and enquiries revealed a public execution in the square outside. Since the hanged man wasn't a good type—a murderer or the like—his family were not permitted to cut down his body, and the poor wretch was left dangling.

Meteors for the Arabs; Meteors for the Israelis. Shades of Sir Basil Zaharoff. The idea of selling military aircraft to both sides disgusted me, but no one took notice of my suggestion that having sold to the Egyptians, Israel's business should be diverted elsewhere.

The Jews added lustre to their reputation as clever bargainers when it came to ferrying their 'planes to Tel Aviv. When we told the Israelis our required fee for ferrying the Meteors, they replied that they could get it done more cheaply, and a London aviation firm wrote saying they would be doing it. They assured me that their man was an excellent pilot who had flown Spitfires to Israel, and had done a conversion course on Meteors, receiving a good R.A.F. report. With sublime cheek, they asked if I would be good enough to write to the Israeli authorities giving their pilot my blessing! They were certain that when I saw and talked to him I would be convinced. He arrived at Gloster's, a pleasant enough young man, but he had never flown Meteors with overload tanks, nor had he made long-distance flights in them. I told him that with much regret I could not give him my blessing.

The ferrying of jets was, indeed, no easy task, and was far more complicated than delivering piston-engined 'planes. To be successful and economic, a tremendous amount of detailed planning and experience was needed.

Dominating everything was the remarkable thirst of jet aircraft. The lower they flew, the more fuel they consumed, and the "long hops" to the Middle East meant flying at least six miles up, preferably nearer eight. Coming down to a thousand feet meant staying up for only half the time (and covering half the distance), so a pilot could not afford to get lost. If he did, and was lucky enough to come down at an aerodrome, he was faced with further problems. The Meteor's engines then ran on paraffin which, in those days, was available only at our arranged stopping places. Even given the paraffin, it required special, very fine, high-pressure filters to remove particles of grit and dirt which might cause engine failure. Oxygen, air, special fittings and pressure connections were necessary, together with high capacity starting batteries sufficiently powerful to whip over the engine fast enough for them to light up and take over as gas turbines. To put down at an unstipulated aerodrome meant being denied all these facilities—plus Customs and monetary difficulties. In short, you were stuck.

Ferrying jets is a job for experienced men who have not only taken "conversion courses" but know their aircraft and equipment inside out. And even though we planned these operations to the nth degree, I still sent the 'planes over in twos, so that double the experience, compasses and radios were available in case of trouble.

An annoyance to me was the difficulty I had borrowing pilots from other firms in the Group for ferrying when I was hard pressed: despite the help I had previously given them by training their pilots in the handling of jets—when we were the only people in the Group to have jets and jet experience. All I received in reply to my requests were excuses, so I went to Rolls-Royce and Blackburn's, who were glad to lend a hand. There was no shortage of unknown volunteers who wanted to ferry 'planes, but too much was at stake to consider employing them. As it turned out, I saw well over a hundred Meteors away, and although there were weather and mechanical delays, none flown by my pilots suffered a scratch.

The annual show at Farnborough, organized by the Society of British Aircraft Constructors (S.B.A.C.), is our aircraft industry's shop window, and unquestionably the greatest display of its kind in the world. It attracts a dazzling array of Service and civil representatives from almost every nation; the public come in their tens of thousands; millions more watch it on T.V.; the Press and newsreels are present in force. Because it is a shop window, the accent is on the spectacular; to dazzle—and often to baffle—the audience with noise, speed and breathtaking performance. The public must be given thrills, and the pilot, rather than the 'plane, receives—and often seeks—the glory. This was typified when one pilot's wife was heard to say: "I do hope my husband makes a better bang today than so-and-so. It can mean so much to us."

The ballyhoo has reached crazy proportions. Rivalry between companies has led to them urging their flyers to try to outdo the pilots of rival companies. This can be seen in the new manœuvres that are devised, and in the way relatively untried 'planes are "pushed". Farnborough, in fact, is no longer aimed at demonstrating an aeroplane's best features to a select, knowledgeable audience, but is, in fact, something of a Roman holiday.

Very often the paint is hardly dry on the prototypes flown at Farnborough. In many cases they have done almost no flying, yet pilots are urged to "put up a good show". I have had a senior (non-flying) technical executive say to me: "We've got to see de Havilland's off today. You can go twenty or thirty miles an hour beyond the limitation." The 'plane referred to was at that time limited to 400 knots, and I replied: "Not likely—unless I've got a written clearance from you to say it's all right." Needless to say, the clearance was not forthcoming.

I have flown new aircraft with controls that were solid at the speeds I was expected to reach. To turn quickly was impossible, as was the rapid pulling out of a dive at speed. In unguarded moments, other pilots confessed to similar problems with their own aircraft, and it is a miracle that there are not mass disasters at Farnborough every year. After one display, "Dutch" Kindelburger, boss of the vast North American company which makes the Sabre, was reported to have said: "It's madness, and asking for trouble."

Displaying an aircraft at Farnborough is difficult enough without adding unnecessary risks. Each 'plane has its own characteristics; its strong points and its weak ones. Customers' requirements must be considered too. If they are interested in ground-attack aircraft, there is no point demonstrating high rates of climb. Instead, the pilot must concentrate on showing the 'plane's steadiness in a dive. Its recovery climb, and its low-level manœuvrability. (The Swiss, for example, operating from small mountain aerodromes, are interested in short landings and take-offs, and tightness of turn.)

The pilot must add some touch of showmanship, of course. A high speed run, low down and noisy, coming suddenly from behind trees or a hangar, can usually be guaranteed to give the crowd a thrill of surprise—especially with breathless commentators whipping it into a froth. Added effect can be given to a high speed run if the engines are suddenly throttled and the air brakes put out to visibly pull-up the 'plane. Conversely, acceleration from a low speed is an impressive sight. Sudden and unexpected manœuvres go down well, such as a steep climbing turn after take-off, with the wheels still retracting.

These things—trying to display the 'planes in a manner to suit every type of customer, and adding touches of "colour"—are difficult enough in brand new aircraft without pilots becoming involved in unrestricted competition.

Although Farnborough is impressive to the layman, there is a lot of sham about it. The chap who only shoots by at high speed may do so simply because his 'plane turns badly at low speed, or lacks manœuvrability. The high speed zoom climb looks magnificent—yet the sustained full power climb at medium speed, while not looking so thrilling, is a truer indication of a 'plane's qualities in a climb.

Since many of the 'planes are seen in a relatively undeveloped state, their performances in public are no indication of their real value as fighting machines. Witness the Swift's impressive high speed passes and supersonic bangs at Farnborough, yet it was a flop as an interceptor fighter. Many of the 'planes displayed are years away from service— and obsolete when they do eventually reach their purchasers. Then there is the question: "How impressive would these aircraft look if foreign 'planes were flying against them?" A misleading impression of performance is added by the aircraft being incomplete: many lack guns, radar and other equipment vital to their real jobs, and for the sake of lightness carry just enough fuel for their few minutes in the air.

Despite all these skeletons in closets, Farnborough is undoubtedly the best organized show of its kind in the world. I have displayed aircraft at numerous shows in many countries, and Farnborough is way above the rest. I recall particularly a well-attended show at Boston: no one knew what was going on, the instructions were verbal and contradictory, and timing and sequence vague.

Much of the supremacy of the S.B.A.C. display is a result of its superb timing and precision. The show is kept alive and full of movement. The flying is in charge of the Group Captain commanding the flying at the Royal Aircraft Establishment, aided by a committee of S.B.A.C. pilots, and I think a large measure of the show's success is due to its flying display being in the hands of airmen.

A schedule is drawn up giving each 'plane its allotted time. If the aircraft have been seen before, are a bit *passé*, or—like airliners—are not to be chucked about, they are put in circuses which merely do a flypast. This also accommodates firms who have nothing really exciting to show, yet are entitled to representation by virtue of being members of the S.B.A.C. Helicopters and light aircraft are sandwiched between other "turns". A pilot is usually allotted five minutes in which to show what he can do, although the heavier 'planes might be allowed eight.

At noon on each day of the display the pilots are given a briefing. The plan of the day's programme is gone over, deletions or amendments are discussed, and the Met boys explain the sort of weather to expect. There is a post mortem of the previous day's flying, and anyone who has misbehaved is slapped down. A pilot who has infringed the rules may be disqualified.

This happened to me in 1953. As a result of the break-up of de Havilland's 110 the previous year, which caused the deaths of twenty-eight spectators, and injury to some sixty more, it was ruled that no 'plane could turn towards the crowd at a distance of less than three miles. In my case I went wrong when the control tower radioed me that I had one minute left in the air. If I turned away from the crowd to get into position for approach and landing I should hold up the next 'plane awaiting take-off at the end of the runway—an awful "black" that would lead to criticism. My natural reaction was to get into position as quickly as possible, and this meant pointing my nose towards the crowd. When I came down I was informed that my flying was "finished". This meant that Farnborough would not see the Javelin, for no one else in the country at that time had any real experience of flying it. But things were smoothed out by the President of the S.B.A.C., and after apologies all round I took the air again.

These precautions may help the crowd (although not necessarily the householders living around the airfield) but they are of no protection to the pilots. For them the basic danger remained—and still remains: the menace of "ropey" prototypes being pushed beyond their capabilities and known limitations for the sake of five minutes of spectacle which proves nothing.

An odd situation arose in connection with the Royal Review Flypast in Coronation year. I received my first intimation of the part I was to play in it by way of a copy of a letter sent to the firm by the Ministry of Supply. This was in April, some three months before the show. The letter was a piece of typical ministerial buck-passing, and said that the Ministry would like to see the Javelin take part "providing it didn't interfere with experimental work. But they shifted the re-sponsibility for a "Yes" or a "No" on to our shoulders.

The R.A.F. were running the review, and they had no doubts in the matter. Almost simultaneous with the M.O.S. letter came an inch-and-a-half thick R.A.F. Operational Order outlining the flypast. In it was the positioning and speed of the Javelin, and the time it would pass the Royal stand. It was also shown in silhouette in the plans of the aircrafts' positions and height.

I went to the boss for a firm decision. He said: "It's silly. It'll muck up our work. We won't take part." None the less, calls came from the R.A.F., and I could only tell them that I didn't think we'd be there. The Air Force persisted, and I went again to the old man, asking: "Can I tell them that your decision is firm?" He replied: "Leave it with me"—and promptly went on holiday. As the weeks dragged by, and rehearsals got under way, the R.A.F. 'phoned me daily. I told them over and over again: "Don't ask me. I only fly the damned aeroplanes. It's up to the management."

But still Gloster's prevaricated. Then we lost one of our Javelins, leav-ing only WT-827, the equipment-testing prototype. She hadn't even flown—and didn't until July 7, eight days prior to the display.

The R.A.F. had meantime turned on the heat. Air Vice-Marshal the Earl of Bandon, who was in charge of the flypast as head of 11 Group, R.A.F., sent a stinker of a letter to the firm. With ill grace it gave way, and I was told that the Javelin would take part. This was a fortnight before the review.

There was an argument as to who should pay the flying expenses, and it was finally agreed that the M.O.S. would foot the bill. Since the Javelin would not be available until unpleasantly close to the day, I went over the course in a Meteor—also paid for by the Ministry.

The size and elaborateness of the R.A.F. Operational Order was worth a knighthood to the senior officer responsible. The amendments alone deserved an O.B.E. The whole thing was thrown on to the pilots; it boiled down to—"follow this route and get yourself past the reviewing stand at such and such a speed at such and such a time."

Its timing arrangements, among other things, caused me considerable amusement. I had but two trial runs in WT-827 over the route. On the first occasion thunder storms scattered rain and rough air, and I went round at a reduced speed to save the new paintwork. Next day a telegram arrived. It said: JAVELIN SIXTEEN SECONDS LATE. . . . DISGRACEFUL. THIS HAPPENING ON REVIEW DAY WILL UPSET THE ENTIRE PROGRAMME. . . . A.O.C. II GROUP. FOLLOWING THE NEXT DRESS REHEARSAL, A FURTHER TELEGRAM CAME: GLOSTERS SIX SECONDS EARLY YESTERDAY. . . . NOT GOOD ENOUGH. . . . A.O.C. II GROUP.

July 15 was clear and sunny—although a bit rough for flying because of turbulence. I passed the reviewing stand at well over 600 m.p.h., hanging on to the controls with both hands and being bumped through the sky. I returned to Moreton Valence, and at about five p.m. this telegram came: magnificent show glosters. . . . wonderfully successful. Actually I had been one-and-a-half seconds early, so I sent back: believe your telegram wrongly addressed. . . . please check. . . . waterton. Next afternoon a harassed Squadron Leader rang my secretary. "The A.O.C. has put me on to checking telegrams to your boss," he said, "but I can't find any mistake." Her giggles gave the game away, and the next time I saw "Paddy" Brandon he said laughingly: "Well, you got your own back, you so and so."

11

CANADA AND THE U.S.A

At a party held at the Dorchester during the winter of 1948/9—I have long since forgotten what was being celebrated—Sir Roy Dobson, one of Avro's and the Group's directors, asked me: "Going to Canada to fly that new fighter for us, Bill?"

I replied: "I suppose I'm as big a sucker as the next one."

Dobson laughed. "Big sucker. That's a good one!" He laughed again and moved on.

I went about my business, the Dorchester encounter forgotten, until October, 1949, when I was hurriedly recalled from Egypt to go to Canada to test the fighter built there by the newest of the Group's firms: Avro-Canada.

There was the inevitable sourness and snarling over my being chosen for the job, but the reason was simply that I was by far the most experienced jet pilot in the Group at the time. I had flown hundreds upon hundreds of hours in jets, single and twin engined, and knew the Rolls-Royce Avons fitted initially in the CF-100, as the Canadian 'plane was called. An added factor may well have been a political one: to boost national pride by enabling the Group to say that Canada's first completely home-produced jet fighter was initially flown by a Canadian-born pilot.

Avro-Canada's own chief test pilot, Don Rogers, was an exceptionally fine flyer, but he then lacked experimental and jet experience.

And so I flew back to Canada in December, 1949, the country I had left eleven years before, and had not visited since I ferried a Boston out of Montreal to Prestwick in 1943. At Toronto's Malton aerodrome I was met by Don Rogers and Edgar Atkin, at the time Avro-Canada's chief engineer. Avro's vast plant is on the north side of the aerodrome, and there I was introduced to the general manager and other senior executives. Within an hour I was in the hangar looking over my new charge. Around the new aeroplane were crowded R.C.A.F. officers and civilians of important mien. I recognized one of the latter as a fellow ex-cadet of the Royal Military College of Canada. We greeted each other, and I said: "What do you do for a living these days, Bud?" "Oh, I'm the Deputy Minister." I looked puzzled. "Of Defence," he added. I had indeed been away a long time!

I liked the new plane—although she was by no means ready for me, despite the frantic cables to Egypt and the hurry to get me across the Atlantic. She was a big 'plane, her twin engines close into the fuselage—exactly as I had wanted Gloster's to do with the Meteor. With her black paint, white markings and tall, stalky undercarriage, the CF-100 had a decidedly Germanic appearance.

I spent a busy day acquainting myself with the Avro set-up and the people with whom I would be working. The Canadians were a refreshing bunch. By English standards they were inclined to be brash, crude and showy—but what a spirit of bustling enthusiasm they showed. They *wanted* to get things done.

While the CF-100 was being prepared for flight, I busied myself learning all I could about her—structure, controls, electrics, hydraulics, fuel system, pneumatics, and so on. I studied wind tunnel reports, calculations of expected performance and aerodynamic behaviour. Some pilots, like car drivers, never much bother to learn what goes on under the bonnet. I always bothered: and to that I attribute being alive today.

I formed a deep respect for Avro-Canada's aerodynamicist. Unlike some others of the breed, he would tell me what I could expect from the 'plane—and was invariably right. He had had a great deal of wind tunnel work carried out both at Farnborough and at the Cornell University tunnel at Buffalo, N.Y. Everything had been tunnel tested: even the wingtip tanks. Such thoroughness was an eye-opener, and gave me great peace of mind.

I flew a great deal with Avro's two pilots in Lancasters, Mitchells and Ansons, to learn the aerodrome layout, the surrounding countryside and the *locale* of other airfields in the province. I ground ran the CF-100's engines, worked her hydraulics, and called for one or two changes in cockpit layout. To my delight—and again in direct contrast to past experience—the alterations were made expeditiously and without argument. I liaised amiably with the civil airport's flying control staff and the R.C.A.F. personnel involved in the project.

The atmosphere was tense, not only at Avro's, but throughout the country. Most of the money financing the fighter came from public funds, and since the North American taxpayers are quick to raise hell, the enterprise had become the centre of a bitter political squabble. Canada had built fighters before, but she hadn't designed and built them, and many people felt that she should stick to the old policy of building proven 'planes under British and American licence. But none of the existing designs quite filled Canada's particular requirements: extremely long range, a good take-off and high rate of climb. The British had the latter, but not the range. The Americans lacked climb performance, and required huge, costly aerodromes—not currently available or immediately practicable in Canada with its limited population of some fourteen millions. The Canadian Air Force wanted their own, specifically designed, aircraft. Further, they were anxious to build up their own aircraft industry in order to make themselves independent, in part, of Britain and America.

As is so often the case in Canada, however, a large section of the population lacked confidence in their country's ability—and, of course, certain British and American manufacturers did not want to see their Canadian market disappear. If the Canadian Jetliner, North America's first jet airliner, and the world's second (it was just pipped by the Comet) and the CF-100 flopped, the R.C.A.F. and the Government would have a lot of answering to do. In no small measure the future of the Canadian aircraft industry rested upon the fighter I was about to fly. Small wonder, then, the state of nerves around Avro's and in Ottawa.

I had no personal doubts about the CF-100's success. She looked right and seemed right. There would certainly be troubles, but teething problems are common to all aircraft. It was clear that I would have to "sell" the 'plane as well as fly her, for the critics and weak-of-faith would not understand the ramifications of experimental work with its inevitable setbacks. Even those in favour of the 'plane expected it to work right off, as could be seen from their sausage-machine production set-up.

There was pressure to get the aircraft in the air before the end of 1949, but it could not be done. Faces dropped, and morale was low. When I saw the aircraft at the beginning of December I doubted whether she'd be ready in four months. But this was Toronto, not Gloucestershire, and on the evening of January 17, 1950, I commenced taxiing trials.

The CF-100 had foot-operated brakes—unlike the hand-operated British type—but I'd had experience with them in German and American 'planes. I found those in Avro's 'plane too light to be operated by clumsy feet, and had the pedal built up and operation made heavier to avoid over-braking. There was wild excitement as the CF-100 moved along under her own power for the first time. She was fitted with nose-wheel steering, and I found it over-sensitive and wrongly geared, causing the 'plane to wander. I disconnected it from the cockpit control. The 'plane then ran smoothly and true, and handled admirably. Although the main wheels were relatively close together, the excellent Dunlop plate brakes turned the aircraft adequately, and with practice this turning could be appreciably helped by the use of the engines. Despite the snow on the runway, the CF-100 was at least the equal of the Meteor in ground handling, and that was good enough for me.

The taxiing trials on 18101 (as the first CF-100 was numbered) went off better than those of any prototype I handled. I beat up the runways, down wind, into wind, and across wind, at increasing speeds, stopping only for examination of the brakes or to allow an airliner to take-off or land. Then I tested control effectiveness at near take-off speeds. I could position the nose where I wished, and the elevators were O.K. The rudder was satisfactory, and pressure on the stick told me the ailerons were responding. The brake plates glowed red in the gathering dusk, due to stopping from high speed. After only one sortie of fifty-five minutes I was satisfied—most unusual for me. There were some minor adjustments to be made, and a thorough inspection was needed, for I had deliberately bounced the 'plane over rough ground to "shake things up". But all in all 18101 seemed an excellent job of work.

January 19 was cold and bright, the wind blowing almost straight down the runway: ideal conditions for a maiden flight. The hangar was an agitated, nervous, excited, confident confusion of activity as final inspections and checks were made. The chief inspector told me the 'plane would be ready after lunch, by which time a party of R.C.A.F. and Government big-wigs would have arrived from Ottawa. The presence of V.I.P.'s meant photographers and newspapermen, a damned nuisance, for that involved answering questions and posing for pictures—the last thing I wanted to do when I was feeling a bit twitchy, wondering if I'd remembered everything, and knowing that despite satisfactory ground trials the 'plane's performance in the air was still an enigmatic question-mark.

Since I had not been informed officially of the coming of the V.I.P.'s, I felt that I was at liberty to dodge both them and their attendant journalists. As an Air Force transport hove in sight, I climbed into an Anson and flew off for a last minute check of runway and wind conditions. By the time I landed it was too late to lunch in the senior staff canteen where the visitors were, and I grabbed a snack in the airport coffee shop. I gathered that the Minister of Defence and the Air Force Chief of Staff wanted to know where Waterton was, and were baffled and narked to learn that he was up in the Anson. By the time they finished lunch I was already in the cockpit of the CF-100, ready to start-up.

The entire airport staffs came out to watch the event, and the road past the airfield was blocked by parked cars. After a double cockpit check, an engine run-up and a brake-holding power test, 18101 was ready for her début. I was carrying half fuel, the 'plane was light, and in the low temperature the Avon engines behaved superbly. Acceleration was tremendous, and in less than 500 yards we lifted cleanly from the runway. I throttled back, and at 140 knots climbed to 500 feet.

I gently braked the wheels and pressed the UP button to raise the undercarriage. But, as on all my first flights in prototypes, part of the machinery had gone wrong—the button did not want to be pressed home. I tried again, more firmly. It still resisted. I shrugged, and decided to leave the wheels down and fly slowly. On that type of undercarriage installation an electrical ground lock was fitted. When the weight of the 'plane was on its legs, to prevent the pilot accidentally raising the undercarriage, a switch was fitted which required forty pounds pressure to work—as against a three to five pound pressure when the 'plane was airborne. When the aircraft took the air, this safety switch should have been automatically released. To override the switch was to ignore that something was wrong. This, I knew, had been done on another aeroplane, resulting in a belly landing. So I left things as they were.

I flew 18101 for forty minutes at up to 5,000 feet and 180 knots, testing airbrakes, flaps, controls, turning, and getting the "feel" of the aircraft's general flying qualities. There was nothing to worry about: she seemed a sound design.

An R.C.A.F. Mitchell took photographs and a film. When I had used two-thirds of my fuel I came in on a straight approach. The 'plane was as steady as a rock, and touched down at about 100 miles an hour within the first 150 yards of runway. We stopped with smooth ease within 600 yards of the start of the runway. For a first flight I was well satisfied. The CF-100 proved pleasant to fly, and take-off and landing were easy.

After I had shut down there were photographs and talk. The brass, sweating it out on the ground, is always rather pathetic at such times: looking at you with spaniel-eyes, pleading to be told the best, terrified they'll hear the worst. For once I was able to be cautiously optimistic.

Six days later, the undercarriage lock fault rectified, I was able to fly 18101 again. The trouble had been a simple, climatic one. The shock-absorbing undercarriage legs were British developed and, unknown to the makers, were contracting slightly in the cold. Consequently, when the 'plane's weight came off the wheels in flight, the legs did not "stretch" as much as they should have done—there was insufficient "stretch" to release the electrically triggered safety switch. When modifications were made the undercarriage gave no further trouble.

On my second flight in the aircraft, I got down to exploring her low speed and stalling characteristics. Nothing to worry about there. But on her third flight we ran into trouble. Quite simply, I "bent" the aeroplane.

Sir Roy Dobson was visiting Canada and wanted to see what the 'plane would do. I did a mild beat-up. Nothing elaborate: just high and low speed flying, with rolls and tight turns.

Previously I had reached 430 knots at 12,000 feet—beyond the point at which a Meteor ran into compressibility trouble. There was no sign of compressibility—but the elevator trim control started to seize-up. It seemed there was distortion somewhere in the back end, and it put an end to ideas of going any faster until investigations were made.

When I landed after the beat-up we discovered that the wing-root fairings were torn and twisted. This was disturbing, for it meant, quite simply, that the wings were bending and twisting far more than they should have done. And this was happening during only mild manœuvres—a fraction of the treatment an operational fighter would have to put up with.

Careful examination of the 'plane revealed that the centre section cut-out under the engines, was "soft". When I flew her again I was permitted only gentle manœuvres while the stress office scratched its head and wondered how to cure the aircraft's depressing defect. During February I made only four flights in the CF-100.

In March, on my eighth flight, I was accompanied by the CF-100's project designer, young John Frost, ex-de Havilland's, and very much the keen English public schoolboy type. Here was another delightful contrast to England, where I was never able to find a designer with spare time enough to fly in his own creation. Frost also came with me to Ottawa to show off the 'plane to the Governor General, Viscount Alexander, and the visiting Prince Bernhardt. The display was a tremendous success despite my "bending" the 'plane for a second time—even though it had been reinforced.

When I was not carrying Avro or Canadian Air Force pilots in the rear cockpit, I was often accompanied by Frank Spink, a ground engineer who watched over the CF-100 like a nursing mother. He was made my "flight engineer" when I first arrived at Avro's—although I pointed out that such an appointment was unnecessary for a fighter and that I would prefer to fly solo. As it transpired, I found Spink most helpful and useful for compiling notes during performance trials. He was broken-hearted when I refused to allow him to accompany me during the CF-100's initial flights, but I could see no point in risking two necks in the untried 'plane. Before my time at Avro's was over, we flew together a lot, and the arrangement worked well.

Putting right the cause of the CF-100's "bending" progressed slowly, for the firm did not want to take risks. Other firms did not mind taking chances with their pilots, for although they could lose all the time in the world on modifications, they expected the flyers to make up for it at the end of the line. The pilot who takes reckless chances and gets away with it is hailed by the chairborne as a helluva "press on" fellow. But when something goes wrong, and he breaks his neck, the same people will call him all the stupid clots under the sun. The sensible pilot insists that things be checked and double-checked—and as a result many faults are often found on the ground which could be more than a little embarrassing if discovered in the air.

When she was back in the air, the CF-100 gave every promise of a fine top performance. Rates of climb, level speeds and Mach numbers seemed to be above expectations. Without elaborate tests and equipment, our figures had to await confirmation, but the 'plane would fly rings round an escorting Vampire, despite the best efforts of that aircraft. The CF-100's qualities showed during a trip to Washington in May. At between 25,000 to 30,000 feet, although well below continuous cruising power, the average speed worked out at 575 m.p.h.—virtually the top speed of the Meteor at its best height, which was considerably lower. From sitting stationary on the runway, the CF-100 reached 40.000 feet in two-thirds of a Meteor's best time, and set up an inter-city record of better than 638 m.p.h. between Toronto and Montreal. Speeds of 660 m.p.h. have been quoted for her, with a 2,000 mile range, and calculations revealed, that I had achieved more than .9 of the speed of sound. Later, a strengthened CF-100 with modified controls exceeded the speed of sound in a steep dive at height.

The purpose of the flight to Washington was to show the 'plane to the Americans. They had evinced great interest in her, and the Canadians, no less than anyone else, were interested in getting their hands on U.S. dollars. I felt privately that if the Americans were sufficiently interested they could have come to Canada instead of Avro's taking the 'plane to Washington. I suspected, further, that the Yanks did not propose buying, anyway, but wanted to sample the Canadian fighter's features and performance for themselves. When they saw the 'plane they did, indeed, go through it with a fine toothcomb—without buying.

When we arrived at Washington, the Americans were, I thought, rather offhand to the Canadian Air Force chiefs. My blood boiled to see the way they were treated as second-rate poor relations, and the Canadians unfortunately lacked that air of down-the-nose patronage with which the seediest, down-at-heel Englishman can successfully squelch that type of ostentatious boor.

There was a slight setback at the beginning of the display when a starting trolley fused and I couldn't start one engine. This was put right, and I aligned the 'plane along the runway—one the Americans did not use for jets because they considered it too short. But the CF-100 did not need a long take-off, and made them duck as it screamed over their heads to start a performance with which no American 'plane of its class could hope to compete. After the beat-up one or two U.S.A.F. Generals were needlessly patronizing, although the lower ranking officers were charming, evincing a great interest in the 'plane—as did Admiral Cassidy and the U.S. Navy flyers present.

By the late spring of 1950 I had established the CF-100's pattern of behaviour and performance, and about the same time a novel and exciting scheme began to take shape. The idea was to fly the CF-100 across the North Atlantic direct to England for the Farnborough Air Show. The principal alterations necessary were the fitting of long-range fuel tanks to the wingtips, and the installation of more comprehensive radio equipment, and since this work was already scheduled it would not be wasted. Sir Roy Dobson was keen on the idea, as were Avro-Canada and a section of the R.C.A.F.

I estimated that by flying from Newfoundland or Labrador, and taking advantage of winds, I should be in London in about three-and-a-half hours, averaging more than 600 m.p.h. On calculations, the CF-100, with internal and tip-tanks, could cover 2,500 miles in still air. The prevailing wind would add a further 300 miles to her range, so there was no technical difficulty against a trip that would earn the fighter world-wide publicity.

I went into the plan carefully. The firm speeded up the tip-tanks. I discussed safety equipment with the R.C.A.F., and Trans-Canada Airlines were most helpful in deciding upon the best radio and routing. We went into the question of spares and facilities, and I talked with Shell about laying down a supply of fuel in Newfoundland. Names of R.C.A.F. navigators were put forward, and the project became well advanced. I returned to England for a fortnight to make arrangements at that end, and to check the route. In this T.C.A. were again very helpful, and I made the crossing virtually as a member of their aircrew. Back in Canada it needed only the installation of the fuel system and a few dummy runs before setting off.

But it was not to be. Ottawa was still starry-eyed at the possibility of selling the CF-100 to the Americans, and there was talk of the U.S.A.F. testing the aircraft at their own tactical establishment in Florida. If this was to be, radar and armament work would have to be carried out before the 'plane flew to the States. So the trans-Atlantic flight fell through. The Americans didn't buy, yet the flight to Farnborough, apart from increasing Canada's aeronautical prestige, would surely have sold 'planes for them, too. Instead, a Canberra later made the trip. The CF-100 would have made the crossing a greater sensation, for she could show her heels to the Canberra.

In July, 1950, the second CF-100—18102—came along. Like the first, she was powered by 6,500-pound-thrust Rolls-Royce Avons, bought by Avro's to power the prototypes. Their own Orenda engines, when they came along, would be fitted in the production fighters. As before, I first flew her solo, then took up Bruce Warren, a new Avro test pilot, in the second seat.

In August I was directed to show off the 'plane at a display in Montreal. The occasion was the visit of the Australian Prime Minister, plus a gathering of Canada's top industrialists. Apart from the CF-100, Canada's first home-built Sabre would be on display.

The trip went down very well. I passed over Malton at 10,000 feet at speed, and gently climbed to 23,000 feet, holding the engines well below their full power. Thirty minutes and ten seconds later I crossed Montreal airport, 320 miles away. I had averaged 638.5 m.p.h., about five to eight of them coming from a tail wind.

The display followed the usual pattern, except that the 'plane was at that time still restricted to a maximum speed of 500 low down. I made up for this by demonstrating the 'plane's low-speed handling, manœuvrability, acceleration, take-off and landing.

From Montreal I was ordered up to Ottawa to fly the Honourable Bruce Claxton, the National Defence Minister, to a meeting in Toronto. I told him I'd get him there in half an hour—and did. Unfortunately, the air conditioning system packed-up, and by the time we arrived we were both kiln dried.

August was highlighted by a flight in which I carried Ronald Keith, editor of *Canadian Aviation*. We covered forty miles in three minutes, forty seconds: some 653 miles an hour. Keith was suitably impressed.

On the last day of the month I returned to America, this time to Boston, where the United States Air Force Association was having its annual reunion. Our bait was General Vandenburg, U.S. Air Force Chief of Staff: we hoped the CF-100 would catch his eye, for he had not seen it at Washington.

I carried Bruce Warren in the rear cockpit, and again the fighter cruised at an average speed of 575 m.p.h. It was hazy over Boston, and

I had difficulty finding Logan airport. Long experience of Britain's industrial haze held me in good stead, however, and we located it and landed without trouble. A R.C.A.F. Dakota brought our ground crew and equipment, and Canada was further represented by her Vampire aerobatic team. Without the Canadian participation there would have been precious little display for the half-million people gathered at Logan. It was the oddest air show I've ever attended.

At no point during its duration did I discover who was running the Boston show. There was no one man with overall responsibility. Instead, arrangements were in the hands of a committee consisting of the Airport Manager, the local Flying Control, the local Civil Aeronautical Authority (C.A.A.) representing Washington, and a couple of U.S.A.F. officers. If the committee had a chairman, I suppose it was the Airport Manager.

It was he who announced that flying would have to fit in with airline operations—and God help anyone who interfered with the normal running of Logan airport. The C.A.A. man, who sported a straw boater, then dropped his bombshell: no aerobatics could be performed over the airport, but must be carried out over the sea. Since the sea was about four miles to the east, this wasn't going to allow the spectators to enjoy much of a display. We could turn over the land, he added, fly over it fast or slow, but must not turn upside-down over it. The Flying Control types told us we would have to pack up and clear off when airliners wanted to get in or out. The U.S.A.F. officers were sympathetic, but could do nothing: it was a civil airport, beyond their jurisdiction.

The leader of the Vampire team was furious; I was nonplussed. Since we were the only people performing aerobatics, no one else was affected by the restrictions. Canada was not going to be able to put up much of a show. We haggled, wangled and argued. Finally, Flight Lieutenant Laubman, leader of the Vampires, and I, hit upon a possible solution. On the north side of the aerodrome was a landlocked bay, ringed by houses. Since it was water, could we do our aerobatics over the bay? The C.A.A. agreed that we could.

Next morning, before the show was due to start, the Airport Manager said to me: "Everything'll be O.K. now. We're getting Shorty in to run the show. He's got lots of experience in running air shows."

Later in the forenoon Shorty arrived. He was in his shirtsleeves and favoured a Panama hat. A pleasant little man, he was manager of an airport in the mid-west, where large air races were held. His main contribution to organizing things, however, seemed confined to an eternally repetitive: "C'm on, folks, we gotta get a li'l co-operation going roun' here." I've often wondered what fee he rated for those inspired words.

The display was somewhat lacking. There was no timing, coherence or sequence. I worked to the time I had been given, and got a clearance from flying control. It appeared that I certainly drew General Vandenburg's attention to the CF-100, for I took off in the middle of his speech. The roar of the Avons cut him short in mid-sentence. The 'plane behaved beautifully, and shattered the Americans, who had never seen a 'plane which coupled such take-off and rate of climb with exceptional manoeuvrability.

Then came the Vampires. They flew as if tied together: an immaculate exhibition. Helicopters flew, VI-type jets were run on the ground, and a B-45 jet bomber staggered off leaving its usual trails of black smoke.

On the second day of flying I had what is known as a "dicey do". It was my custom to fly at about twenty feet from the ground, with flaps partially extended, just staggering along above the stall, and then—in front of the spectators—open the engines to full power and roar upwards in a steep climb. It was both noisy and impressive.

It was in the middle of the operation—when opening up the engines from about 105 m.p.h.—that I was nearly caught out. I had already pulled up the nose, and had the throttles fully opened, when the starboard engine stalled. It's always at such times that things go haywire. The combination of circumstances was the worst possible: at low altitude, the flaps giving high drag, and with virtually no speed. According to every law in the book a prang was inevitable.

Yet I didn't crash. I managed to get the nose down, and gently nursed the aeroplane along until I had sufficient speed to climb away, then, when I had gained sufficient height, I raised the flaps and carried out a normal single-engined circuit and landing. As I had climbed away I had throttled and shut down the rumbling, shaking, stalled starboard engine, otherwise its disintegration would have followed. My actions had been automatic: a reflex developed by long experience.

The acceleration unit of the starboard Avon had gone, and although the Rolls-Royce man in our team made a temporary repair, it still wasn't 100 per cent. right and I refused to fly the CF-100 until it was. A new unit was flown in from Toronto, and we eventually returned to Canada after six days in Boston.

After the show, the "organizers" came along, full of good cheer and this bright item of information: "Now there's been no accidents we can tell you we didn't really care what you did; but we couldn't say that before the show, of course!"

In October I flew CF-100 with new-style air-brakes in slightly altered wings. At the end of the month the 'plane went to America, at their request, for tests at Wright Field. I gathered that the Americans made fourteen flights in the 'plane with thirteen pilots, and flew it against Sabres: a rather unfair comparison between a light day fighter and a heavy, long-range 'plane twice its weight. Although down on speed and Mach number when compared to the Sabre, the CF-100 came off well in manoeuvrability, and outclimbed its rival. But still the Americans did not buy.

Despite modifications, there was still trouble with the CF-100's "soft" centre section, and one incident, during a beat-up at a Toronto air display, was particularly frightening. While pulling up into the vertical climb of a loop, I heard a violent *crack*: a sharp thunderclap of sound clearly audible above the engine and wind noise. Something had gone—but what? I smartly rolled out at the top of the loop, ready to head for open spaces and bale out. Nothing drastic seemed to have occurred, however, for the 'plane flew on without further trouble. But I had the wind up and wasn't taking chances. I cut short the display and came in. The *crack*, we discovered, had been caused by the rupture of metal: the skin of the wing and centre section had again split—and this time worse than ever.

There were other moments of apprehension in the CF-100. On Don Rogers' first flight with me in her there was a sudden, almighty bang followed by a violent gale of wind. I cut speed drastically, ready for Don to bale out while I had a go at getting the 'plane down. But nothing followed the explosive crash, and we saw that the cockpit canopy had gone. Fortunately it had hurtled off without damaging the tail or anyone on the ground, and we landed without trouble. If nothing else, the episode provided us with a free canopy-jettisoning trial.

A later incident was concerned with landing. It was frankly my fault, and I would have chalked up an awful black if the worst had happened. For my error was as fundamentally simple as a car driver parking his car on a steep hill without applying the brake: I forgot to put my wheels down as I came in to land.

The flap controls had been altered in the cockpit, and as I came in I was concentrating so hard on getting the changed procedure right that I neglected to lower my wheels. This would have meant a belly landing had not the control car quickly reminded me of the omission. It demonstrated how over-concentration can be as dangerous as forgetfulness.

On another occasion, when Frank Spink was with me, we just missed what could have been an "unfortunate incident". We were doing low-level fuel consumption checks in marginal weather conditions, when red lights indicated electrical failure: a rarity in the CF-100. Then that which "couldn't happen", did—complete hydraulic failure added to the fun. Drives to both pumps had sheared. Emergency undercarriage flap and wheel brake air systems made possible a landing. Worst of all, elevator control boosters jammed on final approach—but we pulled it off.

I returned to England early in February. What had been planned as a six months' trip to Canada had lasted for fifteen. When I left Toronto, the structural weaknesses of the CF-100 had not yet been overcome, but the 'plane had established an outstanding pattern of behaviour and performance. I was proud to have played some part in the birth of Canada's jet aircraft industry.

Avro-Canada asked me to stay out there, but I felt it would be unfair both to Gloster's and to Avro's Don Rogers. I was soon to discover that my loyalty was sadly misguided.

Work on the CF-100 went on. Today it is the standard night and all-weather fighter of the R.C.A.F. It has been built in hundreds in several different versions all stemming from the two prototypes I flew. A night fighter O.T.U. was the first unit to take the CF-100 into service. They were trainers, which were followed by Mark III's, and later IV's which armed the fighter squadrons. The Americans never took up the aircraft and so the first foreign user is likely to be the reborn German Luftwaffe, for the Canadians have given it squadrons of CF-100's. Canada, having equipped its home units with the fighter, has sent the aeroplane to Europe—one squadron is to complement each of its day fighter wings on the Continent. Prior to this three CF-100's were sent to Britain for tactical trials. They worked in with R.A.F. interception ideas and tied up with British ground radar. They were displayed at the Paris Aero Show and the 1955 Farnborough Air Display—some five years after we hoped to show a prototype there.

The CF-100 has been for many years the West's top night and all-weather fighter. Though new aeroplanes, such as the Javelin, are superior in climb, speed and ceiling, it will still have many years of useful life ahead. No other night fighter has its range and it can undertake night interception and intruder rôles beyond the capability of other fighters which have a drastically limited range in order to gain in rate of climb. Canada can take great credit for the CF-100. It had its setbacks, but no more than many others and less than most. The troubles were mainly of an engineering nature and not aerodynamic, for the CF-100's appearance differs little today from the aeroplane I left in 1956.

12

PILOT AND MANAGEMENT TROUBLES

I returned from Canada to find that all was not well in my department at Gloster's. At the root of the various problems was the oldest of industrial troubles—pay; a headache that was not peculiar to Gloster's, but applied throughout most, though not all, of the industry.

To start with, the industry obtained its test pilots "on the cheap". Almost without exception they came from the R.A.F. or the Navy. By "test pilots" I refer now to the chief and experimental test pilots: the men who flew the prototypes and did experimental work, and who had learned their trades either the hard way, at Service Experimental Units— as I did—or at the Empire Test Pilots' School. There were very few of these men in the country, and they represented only a tiny fraction of the people who called themselves test pilots. The others were "production" pilots, engaged merely in checking and adjusting proven 'planes and equipment coming off the production line. In a firm employing, say, six pilots, only two would be employed on experimental work—

the chief and one other.

Many production pilots were anxious to do this work, for it appealed to their sense of glamour, was far more interesting (and dangerous), and only via experimental work could they hope to become chief test pilots. Since, however, with very few exceptions, they were not qualified to do experimental work—either technically or because they lacked the temperament—their hopes were doomed to frustration from the outset, and because some of them refused to accept the fact, this led, inevitably, to dissatisfaction and jealousies. In consequence, the ranks of the pilots were hardly a unified strength—a situation aggravated by the production pilots' erroneous impression that experimental pilots were paid out of all proportion to the work they did. In fact the reverse was true.

However, as I have stated, the industry in the main did not pay a bent ha'penny for the initial training of their pilots—either experimental or production. That came from the taxpayers' pockets. Nor was it necessary for them to offer large financial inducements in order to get pilots. Young ex-Service flyers, enamoured by the "romance" of becoming test pilots, tripped over one another in their efforts to join firms. Since many managements and technical staffs regarded test pilots as necessary nuisances who, if they did their jobs properly, stood in the way of easy production, these young men played right into their hands. "We aren't worried about pay, We'll play ball just as you want us to. We'll do anything to please you—if only you let us fly."

So they came, "mad keen", and grabbing with stammered thanks and trembling hands the seven-fifty or thousand a year they were offered. Almost invariably a quick marriage followed. Then, in logical turn, children. By this time the pay didn't look so good. The pilots felt they had "positions to keep up". A car and a nanny were often considered essential, and by the time food, clothes and rent were budgeted, that thousand a year wasn't such a crock of gold. Within a short period the keen young enthusiast was a disillusioned grouser for more pay.

And rightly so. For despite his initial wide-eyed impetuousness, the production pilot was doing a vitally important job of work, checking-out equipment costing the country millions of pounds. Since aeroplanes were man-made, they were subject to human error in manufacture and assembly. The production pilot, too, often lost his life. Moreover, Service pilots depended upon him for the correct adjustments and functioning of their new 'planes—the keeping up of standards of performance. He may not have faced the dangers and carried the responsibilities of the experimental pilot, but he did his share.

The pilots wanted more money, they were entitled to more money— and the firms could well afford to open their purse strings. Their annual balance sheets were healthy, and the shareholders never went short of a well-buttered crust.

Every three months or so I would see the management about pay in-creases—including one for myself. For years my salary had been £1,500 a year, but income tax gulped down a third of it, and for this wage I was doing the prototype and experimental work, running two aerodromes, and was in charge of flying control. My number two received £1,350, and the top production pilot £1,200.

I plugged our case incessantly, both verbally and in writing, but I was invariably fobbed off with: "Sorry, we can't do it," or "*I'm* all for it— but so-and-so won't play."

Certain pilots didn't make the job easier. A chap six months with the firm, for example, who had no qualifications beyond a widely shared ability to fly, was furious when I slapped down his mutterings about the absence of firm's cars and other amenities—and there were more like him.

So the management thought I was a thundering nuisance, while some of the pilots accused me of lacking sympathy for their cause. Between the two I had my hands full.

In 1947 the Guild of Air Pilots and Navigators had taken up the torch on our behalf. But, as is so often the case, some of the pilots did not want to be helped. Collating information of the salaries paid by different firms was a tough job. Finally, a table *was* drawn up, and it revealed that Vickers came out head and shoulders above the rest, de Havilland's were not too bad, and that some of the smaller firms such as Saunders-Roe and Fairey Aviation were surprisingly good for their size—the best paid more than £3,000 a year to their top men, and around £2,000 to the ordinary production pilots. At the bottom of the list were our own Group and another (old-established bomber) firm.

The next move came when the Master of the Guild, himself an ex-test pilot, wrote in the politest possible terms to the managing directors of all firms, suggesting that their test pilots were not getting much of a deal and that it would be appreciated if the matter was looked into.

Most of the firms replied—although two of the biggest did not bother to observe the courtesy of even a formal acknowledgement. From the tone of their letters you could tell how they treated their pilots, even if you were unaware of actual salaries. Sir Hew Kilner of Vickers, for instance, wrote saying that he thought his firm looked after their pilots pretty well, but if there was any dissatisfaction, or if the Guild felt things ought to be better, Vickers would be only too pleased to put things right, for they appreciated the important and dangerously vital work their pilots did.

One or two firms responded equally handsomely by raising salaries by as much as a hundred per cent., and introduced pensions schemes by which their pilots would retire at forty or forty-five with upwards of £650 a year.

But nothing was done in our group of companies, and the pilots smarted with resentment. Avro's boys were the worst paid in the Group, and their chief test pilot was literally faced with a strike. It was suggested that all the Group's chief test pilots should meet to talk things over.

We gathered at Kenilworth one day in March, 1952. The time was opportune, we felt, since an industrial court had just *doubled* the wages paid to the Ministry of Supply's civilian pilots. Ordinary pilots were put in the £2,500 a year class, with the chief at £3,200. Adding to our hopes was the fact that the White Paper announcing the awards stated that the M.O.S. pilots did "stooge" flying in the main, while our work was far more dangerous and arduous. We thought we had a strong case, even though we did not have Union backing as did the M.O.S. boys (they were members of the Civil Service Union), nor did we want it if it could be avoided.

We evolved a revised wage scale and drew up conditions of employment which took into consideration a man's experience, the type of work he did, his age, and his duties other than flying. We asked for nothing more than that which the best firms in the industry (such as Vickers) were doing.

We returned to our individual firms, showed our proposals to the other pilots, then handed them in to our respective managements. There was a flurry and scuffle at the Group's head office—but nothing else.

In May we tried again, and convened another meeting. It was at Banbury, and this time all pilots in the Group were invited. Twenty-four out of about thirty turned up, but we were far from unanimous in our voting. One firm's pilots, always the best treated of the Group, with cars and houses provided on a long-term repayment basis, were not keen to "cause trouble".

One of the Group's chief test pilots put his finger on the reason for the impossibility of unanimous action when he said: "I'm scared. I can't stick my neck out any more. I've got a wife and home to think of—*and I can't do anything else but fly*."

There he had it. Poor though the money was for the job, few of the pilots could hope to earn a comparable wage outside flying. They could hurtle through the air in aeroplanes—and that was all. Some had to be pushed even to write short test reports, as I knew from experience.

So at the Banbury meeting there was a great deal of grousing, much acrimony, a lot of snivelling and tearing at throats, but at the end of it all the chief test pilots could only say: "We'll do our best—but that will be precious little since we don't have united support and backing."

At about this time, the late "Mutt" Summers, Vickers' chief test pilot, retired. He was well looked after, as was no more than proper, considering the work he had done. He was interviewed by the B.B.C., and very courageously spoke on behalf of his brother test pilots. Whenever there was a board meeting I was usually invited to lunch with the directors, and such a jollification came along shortly after Summers' broadcast.

One director said he considered Summers "a pretty rum type". Another heatedly protested that Summers' behaviour was "scurrilous after all Vickers had done for him". Then my opinion was asked, and I said that I thought his views courageously frank for a man who had nothing to gain by voicing them.

At later discussions within the firm I argued on, and asked why it was that some firms in the Group were treated differently from others. Why, for instance, were Hawker's pilots the blue-eyed boys, while Avro's were treated so badly? Gloster's pilots, who had flown more 'planes than the rest of the Group's firms put together, were stuck somewhere in the middle.

I was putting my questions to masters of evasion, so I might as well have saved my breath. In June, however, pay increases did come through—an approximate ten per cent. increase as opposed to the hundred per cent. for which we had asked. As a munificent gesture they were back-dated to April 1.

Some of the pilots groaned on—but did little else. I was disgusted by both sides, and my popularity was pretty low all round. Many of the pilots felt I had let them down because I was not prepared to back their petty grievances but wanted to go for the really big issues, while the management thought I should have been supporting the firm's line one hundred per cent.

The only thing that kept me on at Gloster's was my love of aeroplanes and flying. The set-up had precious little else to offer.

A by-product of the firm's parsimony was a dangerous trend by certain pilots to seek fame and fortune by other means—a by-product, too, of the fact that lack of qualifications prevented them from ever reaching the top of their profession. This manifested itself by pilots practising aerobatics when they should have been getting aeroplanes to an "off test" state. When I had one youngster on the carpet because of this, he sullenly whined: "A pilot's got to get himself a public. If you become well known and establish yourself, you're made." He might have been, for some people can, indeed, survive—even flourish—on personal publicity. For my own part I found it a poor substitute for hard cash.

On top of my other troubles I suddenly found myself at odds with, of all people, a photographer. Years before I had agitated for photography as an excellent means of showing designers what happened to their creations in the air. The camera could reveal all manner of aerodynamic and engineering phenomena undergone by an aircraft in flight. It took me a long time to win my point, but a ciné camera was finally purchased.

On my return from overseas, however, I found that the camera was almost fully monopolized by the publicity department. They had placed it in the hands of a photographer, and he was amiably occupying himself by snapping pretty pictures by the score. You could hardly spit without hitting shots of Meteors taken from all conceivable angles, together with charming views of Cotswold villages, the Severn, the Welsh hills—pictures of everything except what was wanted: skin distortion, flaps sucking out and suchlike.

I learned, too, that to take his happy snaps the photographer was flying in the rear seat of a trainer on an average of three times a week—and this at a time when I was trying desperately to cut down the number of unnecessary flights.

I had the camera reverted back to the scientific rôle for which it had been intended—and that caused further pouting among some of the pilots. For they did not see their names in the local papers so frequently in captions that announced: "An unusual view of a Meteor piloted by . . ." It did, however, make me more popular with the resident Government inspectors who had to justify, or render an account for, the number of flights and hours.

The squashing of both the "I must get myself a public" and personal publicity schemes added to my unpopularity, but I argued that my job was to get aeroplanes through as quickly and cheaply as possible, not to win friends.

I don't know what it is about flying that affects young men in the way it does. Perhaps it is the cumulative effect of high altitudes. I have often marvelled at the adolescent antics of pilots and ex-pilots in pubs as they crouch forward, arms extended, imitating the sound of roaring engines or the chatter of machine-guns. I've seen soldiers and sailors inspired by liquor, but I've yet to hear a subaltern mimic the rumble of a tank, or the roar of heavy artillery; or a submariner make mock hydroplanes of his hands, or pretend to be a torpedo. (I am equally amused at the way the "most modest" of heroes always succeed in having the most pictures of themselves taken, the most words written about them, and the most mentions in the Honours Lists.)

Of the young men in the pubs, perhaps the answer is that since their work is really much duller than they would have their audiences believe, it is necessary for them to indulge in exhibitionistic displays in an effort to conceal the fact. I only know that most pilots who have really had "incidents" have never been keen to relive them as saloon-bar skylarks.

My log-book reveals seven severe crashes, one to each seven hundred flying hours, and thirty "incidents", any one of which could have developed into a catastrophe. Some were serious, others have been half-forgotten under the accumulated trivia of time, a few were comic.

On the average something went wrong every 135 hours—about once every five months during my flying career. For years in the Service I went crash-free, and the prang average would be appreciably higher on just the flying I did at Gloster's. On two occasions the responsibility was partially mine: once as a pupil I bent an axle landing; the other time when I came down on the banks of the Seine. For the rest the burden must lie upon bad assembly, bad materials, bad design and bad servicing. Pilots make their share of mistakes, but if I had made as many boobs as the designers and works' people I would have killed myself at least once a year. You can improvise for a 'plane's queer behaviour, or for ropey aerodynamics—but you can't do anything when bits fall off in mid-air.

In this respect I recall Tangmere in 1945, when I was doing gunnery trials over Pagham Harbour in a Mark III Mustang. . . .

A British modification replaced the American fold-over cockpit canopy with a single Perspex bubble. It provided excellent visibility, and extended down to about level with the pilot's thigh. It was attached to a chain, by which it could be wound back and forth on guide rails from fully open to closed, where it battened against the rubber ceiling arc of the windscreen casting.

On this particular morning I was doing steep diving attacks. Starting from 10,000 feet, I would roll over into a dive, putting the nose down to an angle of sixty degrees. The ground rushed towards us as speed built up, for the Mustang was much faster in the dive than the Spitfire.

During one run I glimpsed 5,000 feet flash across the altimeter dial as we plummeted downwards. My speed was approaching 515 m.p.h.— not bad in those days. Then, suddenly, came a deafening explosion, and I was hit by a terrific gale. Its force blinded me and nearly knocked me unconscious. Although I was dazed and confused, I instinctively closed the throttle and reefed back with both hands on the speed-heavy stick.

I remembered stories of Spitfires disintegrating in the air, their pilots finding themselves attached to their seats—but nothing else. I wondered vaguely if this had happened to me.

But no. The gale gradually ceased and I was able to force open my eyes. I was in a 200 m.p.h. climb at 6,000 feet. My plastic bubble had exploded. All that was left of it were pieces of Perspex sticking in my face and to the sides of the cockpit. My goggles were gone, torn from my flying helmet which was ripped at the back.

It had all happened in about twenty seconds, and I have never over-come my wonder at the incredible speed with which these aerial inci-dents and disasters occur. (I remember seeing a flak-wounded Hurricane thirty feet away from me over Dunkirk. Absolutely mesmerized, I watched as the 'plane shed its cowlings and wings as it went spinning down into the sea. It all seemed to happen in nightmarish slow motion, yet the entire disaster occupied no more than a few seconds.)

A far more serious pre-Gloster incident occurred in a Typhoon. The "Tiffy" was a bit of a cow to fly at any time, and was even more so when they slung rockets under her wings. Two 1,000-pound bombs were not so bad: they were streamlined. But it is virtually impossible to streamline rockets, each carrying a five-inch head and three-inch motor, secured to old-fashioned rails hung a few inches beneath the wings. At first eight rockets were fitted—four to each wing. Then some bright spark decided that the "Tiffy" could carry eight more, slung underneath the others. The result made the Typhoon look as though she had barn doors slung under her wings, while a head-on view suggested strings of onions. The bottom row of rockets was about fourteen inches clear of the ground, and not being laterally supported, the entire contraption flapped from side to side.

I had the job of trying it out for the first time.

Initially we carried dummy rockets, so that if there was a prang no explosives would be involved. For safety reasons they were wire-locked to the aeroplane. (This also meant that if anything went wrong in the air they could not be jettisoned to lighten the 'plane's weight.) The first trials went off without incident, although the "Tiffy" waddled along for an inordinate time before reaching the 140 m.p.h. necessary for her take-off with this load, and blundered through the air at 100 m.p.h. below her normal maximum.

On the approach-in one had to be careful to avoid an excessive rate of "sink", and the actual landing had to be gentle so as not to strain the highly loaded undercarriage legs. On the other hand, there was no trouble about "bouncing"—once she was down she stayed down, for with sixteen rockets aboard, the Typhoon had as much bounce as a ton of wet suet.

Next came attack, firing and sight-setting trials with live rockets aboard. Everything was happy until one day when I reached 100 m.p.h. on take-off there was an explosion—and we swung twenty degrees to starboard.

This was anything but funny, for ahead were Wittering's great, massive pre-war concrete and steel hangars. There was no time for detailed thought: the aircraft was not likely to become airborne in time to clear the hangars.

Not the sort of prang to be recommended, I thought. *I must get stopped.*

I automatically chopped the throttle, applied the port rudder and the brake. God, those hangars are approaching quickly. . . . A glum prospect!

We slowed down—and then the port brake started to pack up. *Probably over-heating. . . . Whatever it is, we're swinging to starboard. . . . What's that? . . . The starboard wheel rim is digging in, swivelling us round it. There's nothing I can do save cut the magnetos, switch off the fuel and hope . . . I'll be lucky if we hold together with these side loads....*

Then the starboard leg collapsed, and the wing settled with a grinding, tearing gasp. Clods of earth were thrown up, and I was flung heavily against my straps.

Then came that absolute, uncanny silence, which is always so startling on such occasions because of its contradistinction to what has gone before.

I sat stunned, unable to move. But not for long, for with burst tanks, a hot engine, and damaged electrical leads, a split second can mean the difference between life and death.

I slapped free the quick-release of the safety harness, clambered up and over the cockpit in a flash, and scrambled out of my parachute. With live rockets about I knew that in the event of a fire Wittering would not be a very healthy spot. Such toys are meant for one's enemies, not comrades in arms. As the ambulance and fire tender rolled up at an agonized speed of 30 m.p.h., I was unplugging the last of the rocket heads still attached to the 'plane. The "Tiffy", never a handsome sight, was in a very unhealthy state indeed....

I was in no way criticized for pranging the Typhoon—although I had earned extreme displeasure during my first week at Wittering. I dented in no uncertain manner the Unit's pride and joy: a newly-arrived Tempest V fitted with a Sabre engine.

The acting Flight Commander briefed me on her, and showed me round the cockpit. My flight would not only serve to give me experience of the 'plane, but would also help assess the Tempest as a sighting and aiming platform—to see how steady she was, and how easily and accurately a target could be held.

I signed the entry in the authorization book, and learned that my sighting was to be done on bombers from a nearby Operational Training Unit. I was assured that the Wellingtons and Stirlings welcomed an opportunity for exercising their rear gunners.

After gunning odd aircraft over most of Lincolnshire, I "attacked" a Wellington at 2,000 feet near Market Harborough. My main tanks showed about five gallons, so I switched over to the alternates before going in. I attacked from just above level and from the starboard quarter, then broke downwards under the Wellington's tail with negative G, skidding violently to throw off the rear gunner's aim.

The Sabre engine spluttered—and wouldn't pick up. I checked the fuel cocks. They were O.K. I tried the hand priming pump. It gave a few reluctant coughs. You could keep a Merlin's twelve cylinders going by furious pumping, but not the twenty-four of a Sabre. It was no use. I continued to lose speed and height—and there wasn't much of either left by this time.

I spotted a glider-training aerodrome to the left, made my hiccoughing way towards it, and came in over a road and some trees to land on an out of use runway. Unfortunately, however, I had too little of everything, and pancaked gently through the tops of some trees and a hedge before bumping heavily on to the runway. Vision did not seem to be what it was—there was too much sky, and insufficient ground. This did not turn out to be surprising when I discovered that a water pipe on the ground had torn off the rear wheel, and the Tempest was dragging her rear fuselage and rudder along the deck.

We did not stay on the runway long, but careered off into some ploughed land to the right. I saw a tractor flash past the wingtip—then glimpsed its driver literally fling himself from his seat and scuttle away.

The Tempest was in a bit of a mess. Apart from damage to the rear fuselage and rudder, every twig and branch had left its mark on the wings. The front edges were serrated like much battered bread-knives. The open wheel wells were full of kindling, and from the radiator cooler and air-intake I could hear voices! Inside I discovered a bird's nest, with three young ones plaintively wondering what the devil it was all about. I found them another tree.

After some considerable time, the Americans who controlled the aerodrome realized they had a strange visitor. They seemed to think I was a Jerry until an R.A.F. liaison officer came on the scene. I telephoned my base and they sent a 'plane for me. Its pilot was white-lipped: he had authorized the flight.

They tried hard to get me for that prang. I wrote an honest, factual accident report. I said in it that I was attacking a Wellington from the O.T.U. when my engine cut out, and went on to describe the details. Air Ministry has an Accidents Branch which investigates crashes, and goes through such reports looking for victims. I seemed a dead cert.

A fortnight later the Adjutant called for me and said: "You're for it over that prang. Accidents Branch have sent this." He handed me a signal which stated: "Disciplinary action will be taken against Ft./Lt. Waterton, W. A., as a result of an accident involving Tempest JN 757. It is noted that he was attacking another aircraft without prior authority. This is contrary to section so-and-so which must be incorporated in the Unit Order Book and signed as having been read and understood."

The Wingco was very decent about it. "I'm sorry about this, Waterton, but I'm afraid you're for the high jump. It's partly my fault. I should have noticed what you put in your report and sent it back to you for alteration. But it's too late now. You'll get details in due course."

I tackled the Adjutant, and asked what was likely to happen. A cheerful type, he replied: "Oh, a Court Martial. Or the A.O.C. might deal with you summarily and hand out a reprimand and a log-book endorsement."

I said hotly: "I don't want my log-book endorsed: not that way. I'll stick out for a Court Martial, I think. I only carried out what was authorized to do in the flight book. Besides, you've had the Unit Order Book on that investigation into the airman killed by a 'Tiffy' propeller. I've not seen nor signed the order. Frankly I don't think there's a case against me."

He winced, but said nothing. A week later the Wingco had me in and said with a sheepish grin: "I don't think we'll hear any more about it. Forget it." And I didn't hear anything more, either.

13

THE JAVELIN STORY

Any complaints I had at Gloster's were outweighed for a long time by my love of flying. After my return from Canada, however, the scales began to tilt the other way. Many things contributed to this—lack of discipline (highly dangerous in flying matters at any time); tale-bearing; "playing politics"; complaints from the R.A.F. which proved that standards had deteriorated; a backlog of 160 Meteors requiring testing; a mountainous, untouched pile of administrative work; and a "Bolshie" crew resulting from a feeling of insecurity which I, certainly, was unable to alter, despite continual efforts to do so.

Adding to all the unpleasantness, I found myself frustrated in my efforts to develop the experimental side of our flying. It had long been my regret that Meteor production had always been well in advance of experimental work, for it meant that if aeroplanes had deficiencies they often went into service before we were able to rectify them. We were aware of the faults, but so slow was the work of modification that it was often years before a known defect was put right in service.

The Javelin—G.A.5 as she was first called—was a radical aeroplane.

She had triangular delta wings, and her two Sapphire-type engines, situated at the sides of the fuselage, made her the most powerful fighter in the world at that time. Other triangular-winged 'planes had flown, but only the Javelin carried a delta-shaped tail fitted horizontally atop her great raked fin. The tail, which had conventional elevators, was of the all-moving trim type for steadiness in the trans-sonic speed range and it meant that she could carry normal landing flaps—unlike the tailless deltas. The Javelin was an important aeroplane: on her depended the future night and bad weather defence of the country when other fighters were grounded. As George Carter, our chief designer, projected her, she was a masterpiece. I fell for her when she was no more than an inspired doodle on a bit of paper, and I lived with her through drawings to wooden "mock-ups" to the finished article.

Having experienced weaknesses in machines, I was anxious to correct any flaws—or as many as could be spotted—before the Javelin went into production. I might have saved myself the trouble, for even when the 'plane was little more than a design it became obvious that my complaints were a waste of breath.

I protested, for example, at the quantity of fuel the Javelin was to carry, and the high side rails of the cockpit and windscreen which drastically reduced vision. Although the aircraft was still in the "mock-up" stage, with only a wooden replica produced, endless excuses were given as to why alterations could not be made to these items—excuses that were always a defence of the design office and the works. Never, at any point during my years with the aircraft industry, did I hear: "Such-and-such would be a good thing for the pilot and crews; let's give it to them." It was invariably, albeit silently: "Let's take the easiest way out."

I was horrified when I saw the type of open shrouded controls with low "boost" (hydraulic assistance) planned for the Javelin. The Meteor's controls were always one of that aircraft's poorer points, and those installed in E1/44 (the Gormless) were even worse: they had a heavying-up which made it beyond the pilot's strength to manœuvre the 'plane rapidly at anything above half speed. Yet the Javelin was to have exactly the same controls as those fitted to the Gormless. Again my remonstrance went unheeded.

When the Javelin flew I was the least surprised of anyone that its controls were hopeless. They had no positivity, and were virtually immovable at more than half speed, even with two hands.

The first prototype, WD 804, was ready in late October, 1951. During taxiing trials she showed an alarming tendency to rear-up suddenly at low speed when her flaps were down. After much haggling, I managed to get the trim range of the tailplane altered to cope with this phenomenon, and towards the end of November I took her aloft for the first time. I had carried out much high speed taxiing and I had few qualms as I lifted her off the runway. But I was no sooner well clear of the ground, and was picking up a bit of speed, when buffeting and banging set in somewhere in the tail end. At 200 m.p.h. the whole airframe shook violently—a matter which afforded me considerable concern. As speed was reduced it eased off and I flew around at a very sedate 150 having a quiet look at things. After half an hour I came in to an easy and uneventful landing.

From then on it was nibble here and nibble there at that, and the many other deficiencies which revealed themselves. Never was there any keenness to take the bull by the horns and face up to troubles. The Javelin was easy to fly, had an excellent performance, and showed great promise. She had some dangerous tendencies too, such as reversing her longitudinal control (the stick had to be pushed instead of pulled) near the stall, tightening into the turn, and pitching strongly nose-up when the flaps were extended. In these, and a host of other control, stability and engineering problems which I reported, only lukewarm interest was shown. The design staff—now without George Carter, who had retired out to pasture—treated her as casually as if she were an old, well known Meteor with years of knowledge behind her. The Javelin was their baby, and like doting parents, they blinded themselves to its faults—and took most unkindly to anything which might reflect on their parentage.

After six months' flying the pattern of the aeroplane's performance and behaviour was pretty well established. In the first few months of her life she kept going and, for a prototype, did at lot of flying. What was required was action to face up to her problems and troubles.

One day, not long after the Javelin's first flight, we had some important Ministry of Supply visitors whom we wished to impress. Carefully keeping the speed low in order to move the controls for aerobatics, I performed zooms, upward rolls, rolls off the top, and reversed turns, all in quick succession. I did not loop either then, or at the two Farnborough shows at which I flew the Javelin, for I should never have had the strength to pull it out of a steep dive. After the display a senior member of the design office said to me acidly: "Don't know what you're criticizing those controls about. If you can throw it around like that there's not a thing wrong with them." When you've had to resort to tail plane trim, had both hands on the stick, avoided steep dives, kept the speed down, and sweat copiously, how do you start to argue with such a man?

One of my greatest fears was that when at high speed another aircraft might get in my way while I was in a Javelin. For if the other 'plane's pilot did not take action to avoid a collision, it was doubtful whether I'd be able to.

There is nothing more infuriating and humiliating than to attempt to put over one's points to someone who just doesn't want to know—especially when he is the person who makes the decisions. I am not alone in thinking that this is one of the many serious problems which must be solved if Britain will ever again produce aircraft that really work.

Today the industry is run by two classes of people: accountants and long-haired boffins. The former are interested in making the maximum profits in the shortest possible time, with the least possible effort and outlay. The latter got their wedge in during the war. They have by now consolidated their empire, and with their graphs and figures the boffins can prove anything. Neither accountants nor boffins fly their firm's creations, and in my experience have always shown the greatest aversion to doing so.

The practical engineer who can put a nut to a bolt has been pushed out, together with the test pilot, who is now little more than a stooge for the sales, publicity, commercial and design departments. Yet, ironically, the test pilot is the only person of them all who really represents the user. I have always contended: "As a test pilot you can't shoot a line about an aircraft. If you do, you'll get caught out as surely as God made little green apples. The next pilot who flies the 'plane will discover as much, or more, about it than you have."

'planes are built to be flown by pilots, and are cleared for service by them. How, then, can a firm ignore the criticisms and warnings of its pilots, yet hope to get the 'planes past others? But that is precisely what they try to do.

As a test pilot, with time and the number of prototypes at your disposal limited, you can only hope to hand a reasonably safe aeroplane to the user. You might miss some minor point. Different methods of usage and varying standards of maintenance can cause new troubles. Further: you fly new aeroplanes and cannot legislate for any deterioration that might occur as the hours pile up after long service.

Such things are beyond the test pilot's control—but the *obvious* faults he points out should not go unheeded. Yet they do—for reasons of politics, production and business. And the sufferers are invariably those who fly: the R.A.F. crew, the civil crew and passengers—and the taxpayer.

I have always tried to get an aeroplane to a state in which an ordinary, inexperienced junior squadron pilot would feel at home. I remembered my earliest flying days and recalled how little I really knew about flying and aeroplanes. Joe Prune, and not Bill Waterton with 5,000 flying hours behind him, was the chap for whom the 'plane was built. But what was the good of such idealism when people ignored warnings and let faults go uncorrected?

There are only three ways to stick a job: you love it and don't care about money; you get a sufficiently good screw to justify work you detest; or there's nothing else you can do. In my case I certainly wasn't getting a Gulbenkian income, and my love of the air was rapidly being knocked out of me. So the point came when I decided that irrespective of whether or not I could get another job, my self-respect demanded that I should get out.

In April, 1952, I wrote my resignation, couched in no uncertain terms. The management were most affable. "Why did you write this?" I was asked. "Never write down what you can talk over. . . . Anyway, don't worry. We'll look after you. You can trust us. . . . "

I went away feeling far from convinced.

Two months later, however, another kind of incident disrupted my plans—I nearly "bought it" while flying the Javelin.

"Six days shalt thou labour and on the seventh shalt thou rest," says the Bible—but that was written before overtime and double pay for Sundays were invented. During my seven-and-a-half years at Gloster's, I spent about two hundred weekends at work when I was officially off-duty, and one such Sunday—June 2, 1952—saw me at Moreton Valence ready to fly the Javelin prototype, WD 804.

She had completed ninety-eight flights, of which I had done eighty-four, including the first thirty. After sixty hours flying, she was grounded for a second thirty-hour inspection. (Aeroplanes also have a regular series of daily, between flight major and minor inspections. Additionally, prototypes spend much of their time on the ground for changes, alterations, modifications and adjustments.)

I didn't go much on WD 804 after her second thirty-hour check-up. They had altered her centre of gravity without informing me of the fact—or recording the change on the flight forms—and the hydraulically assisted controls had been messed about with so that their "feel" was different.

On her first flight after the check-up WD 804 proved nose-heavy on take-off, and elevator control at low speeds was bad. Not surprising with a forward-shifted centre of gravity. I raised hell—and gathered black looks from those whose oversight might have broken my neck. Adjustments were made, but flying progressed slowly. It was essential, however, that the Javelin should fly well and fast and give of her best, for the firm had a lot at stake. Time was spent on further modifications, and I was now being pushed by the technicians to keep the Javelin in the air.

Prompting their exhortations was the annual tactical conference of Fighter Command on June 4. Here, among other 'planes displayed, would be our great rival, the de Havilland 110 flown by John Derry.

Although we had our contract to produce Javelins, de Havillands were keen to see us off—and understandably so. (In point of fact, however, we had the measure of the 110, despite the sort of problems which restricted high speed manœuvring, for I had met it in mock combat at 35,000 feet.) For once, the firm's "pushing" was infectious, and I was equally anxious to impress the R.A.F. with our revolutionary-designed mount, having spent so many years with now obsolete straight-winged Meteors.

It was hot, that Sunday, with a copper sun blazing down from skies of azure blue. After take-off I climbed gently to 3,000 feet, checking the controls and accustoming myself to their new feel. Things seemed as satisfactory as on previous occasions since inspection: no more, no less. I moved on to higher speeds.

When flying experimental 'planes I have always tried to keep in close proximity to aerodromes in case I needed to get down in a hurry. On this occasion I pointed the nose towards Oxford and the vast American bases at Fairford and Brize Norton. I thought, too, that it would do no harm to show our American cousins that we also had new shapes on the way.

Still at 3,000 feet, I gently pushed the throttles open, and as the revs increased the willing Sapphire engines tightened the traces. The Javelin really jumps to it when the throttles are belted, and my air-speed rose well above that which any Meteor had ever reached, and there was still a lot to come.

Cirencester passed to starboard, then Fairford and Oxford loomed ahead. I throttled back, and we climbed gently over Oxford before making a 180-degree turn back towards the west. Apart from a few negligible air bumps, there was nothing to report.

The extreme heat of the day emphasized the cockpit's hot, oily smell. It grew worse, and I realized that the cooling system was not working. My clothes soon became sopping wet with sweat. As I turned towards Gloucester I thought how nice it would be if I could spend the rest of the day on the river in the motor-cruiser which now served as my home. But I knew my lot that afternoon would be the writing-up of a detailed report. I eased down the nose, and prepared for a fast run back.

Brize Norton and Witney lay ahead, and I aimed between them. At 3,000 feet the engines were opened to ninety per cent. of their full power, then steadily increased to "full bore" as the wall of air ahead of us mounted up like a ship's bow wave, resisting our progress. We had reached a speed which was already high, but still 30 m.p.h. less than had been achieved in the Javelin, when I felt a few small bumps. They were not serious, but were unusual after the previous smoothness. I put them down to local air turbulence, which was not uncommon on a hot day, and decided there was nothing to worry about.

Then, without warning, WD 804 turned herself into a sort of crazy pneumatic drill. Landscape, instruments—every-thing—fuzzed into a blur. My scrambled brain registered one horrifying word: "flutter"—that sympathy a thing has with a certain frequency of vibration; that resonance which can wreck the biggest bridge if troops fail to break step over it.

Before I had time to close the throttles or open the air brakes, there were two explosive cracks. Then an uncanny, ominous silence as the rattling ceased. The nose pointed itself downward towards the ground which was only seconds away. Something had let go, something in the vital, pitch-controlling elevator circuit—perhaps the elevators themselves. I had a vision of the whole tail having broken off. The stick confirmed my assessment: at that speed normally almost solid, it offered no resistance to forward and backward movement.

Yet everything else was smooth: no concussions, shaking, or banging, and the Sapphires hummed contentedly. As the records later showed, the whole episode had taken two-and-a-half seconds.

I had throttled back instinctively, and the nose now dropped even further towards the ground. Knowing that increased power was destabilizing (causing the nose to rise) I opened the throttles again and was able to arrest the earthwards dive. My hand moved towards the airbrake lever—but I held back. I wanted to slow down, but not too quickly, for to subject a damaged aeroplane to further sudden stress would be asking for trouble.

My immediate impulse was to get out. Yet I knew that no one had bailed-out at such a high speed and lived to tell of it. So I restrained myself, and considered the faint possibility of getting higher and slower. There were one or two things worth trying. I could alter the plane's flight path by varying engine power—perhaps by air brakes and flaps, too, when speed was low. If only the circuit or elevators had gone, perhaps the tailplane still could be moved by the trimmer. On the other hand any of these actions might upset the 'plane's equilibrium. But I had to do something . . .

With one hand on the canopy jettison handle, the stick between my knees to keep the 'plane level laterally, I gently inched the trim wheel back with my left hand. It worked. The nose rose. We started to climb, gradually the speed fell off, and I eased the throttles back gently. At 10,000 feet I levelled out doing 300 knots. Now getting out was a practical proposition.

I switched my radio to "transmit" and spoke to my controller, Roy Julyan, at Moreton Valence. "I've had a spot of bother, Roy. No elevator control. The stick is no longer connected to anything fore and aft. The elevators went between Brize Norton and Witney."

His voice came back, dull and lifeless. "Can I do anything?"

"I'll let you know. I've got up to ten thousand and I'm now heading for the Bristol Channel where I propose to dump the lot in the drink."

That, indeed, was my plan, for no one with a conscience could abandon several tons of explosive fuel and metal where it might fall on a crowded road, town, village or house. But I could not stay up for ever. The engines were thirsty brutes and had already drunk deeply into the Javelin's fuel.

But as I flew westwards I began to toy with the idea of achieving the seemingly impossible—of *landing* the Javelin. At 10,000 feet there is room to play about, so I explored my makeshift handling of the 'plane at circuit and landing speeds. I was able to fly with undercarriage down, and flaps partly lowered, as for landing. I found that I could keep the Javelin to within two hundred feet of any height I wished, and that if I were to drop too low I could put myself in a climb by applying power.

Why not have a try at landing the Javelin at Boscombe Down on Salisbury Plain? Its long runways and clear approaches offered the best possible conditions, and if it didn't work I could "hang around" until I was nearly out of fuel, then abandon the aircraft over the Plain itself. In its sparsely populated area I felt one could leave the aeroplane with little chance of it doing much damage or harming anyone. I told the control tower of my decision and asked them to alert both Boscombe Down and all other aerodromes *en route*.

It seems straightforward enough when put down on paper, but it does not convey—and no amount of writing *could* convey—the tension I felt, superimposed upon a clear-cut conception of what I wanted to do and what I knew to be necessary. My primitive instinct of self-preservation was urging me: *Get out, you fool; get out of the 'plane*, while my reason was saying: *You must try to save the aircraft*. It was a mad, incoherent period of time, with emotion battling against reason—a shapeless mixup.

As I flew on, some sadistic trick of memory brought to mind scenes from my flying past. Had the time come for someone to perform what had often been my job over the years? Would I be identifiable, or would I look like the charred mess of a one-time room mate—the leather of his R.A.F. helmet melted into the black of what had been his handsome boyish face? Would my hands and feet be scarab-smooth claws, or would a dog lap at my spattered brains? I had once put a boot to the tail of one I found eating an instructor beside his wrecked Harvard. Would my bulk be concertinaed into a grotesque two-feet-six dwarf like the one I had identified on a slab? Huge, swollen hands and feet were stuck at grotesque angles to its tiny arms and legs. Or, would I ... *What the hell— I'll have a go. If it's in the cards—that's it. ...*

At last Boscombe Down hove in sight. I called the tower and asked the controller to get his fire tender ready. With wheels and part flap left down, I gradually went down to fifteen hundred feet, keeping my speed around the 200 mark. A wide circuit with the gentlest of turns, and I was aligned heading west towards the main runway. Now to reduce height and speed.

Holding her laterally level with the stick between my knees to prevent turning over, my left hand was on the trimming wheel, my right stretched across my body to work the throttle. Now I was down to a thousand feet . . . now down to five hundred, with my speed 60 m.p.h. fast, for below this I found she lost steadiness and what little control I had. I would have to put her down at this, two thirds above her normal touch-down speed, and forty-five miles per hour more than her normal take-off speed. At this speed she would fly herself off the ground without help from the pilot, but if I could catch her on the ground with the wheel brakes I might pitch her forward and keep her down at the small risk of blown tyres and burned-out brakes.

The aerodrome boundary passed, and she went in as steady as one could wish. If only a gust of wind didn't hit us. The runway loomed up. This was going to be all right; I was going to make it. My gamble was coming off. As I eased back the throttles she touched the runway with but the slightest bump—with smoothness the crux of the whole matter, I had pulled off a "daisy cutter". Yet a split second later all was lost.

Whether it was a fickle variation of wind, or an undulation in the runway, I shall never know. The Javelin, still very much a flying machine, was reluctant to exchange air for earth at that speed. She bounded into the air again to drop gently on to the runway—and to be flung aloft again by her tough, springy undercarriage, aided by her great buoyant wings. With the nose pointing dangerously high it was fatal to open up and attempt to go round. I tried to catch her on the partially effective tailplane, but it was quite hopeless: the rate of movement was much too slow. I was able to do nothing but sit it out.

I could only hope that a leg might collapse, causing the drag of the ground wing to brake, swing and spin her round to a halt. But no: George Dowty built his undercarriages too well for that. In a succession of ever-increasing bunny-hops we bounced along the runway, higher and slower, higher and slower—with a heavier bang every time we grounded. Onlookers said the Javelin would easily have cleared a hangar. Inside it was rough and frightening, sitting waiting for something to go. It soon did.

We dropped from a near-stall with an almighty crash. This was it—a dull, heavy boom, the smell of paraffin, and a sheet of flame and black smoke slashed over the cockpit. Fuel tanks had ruptured and were exploding. This time she stayed down for a bit. The port leg, breaking away from its very mountings, had been driven up through the wing.

The 'plane lurched over, her port wing-tip dragging along the concrete, swinging the Javelin that way and off the runway. I was thrown to the right of the cockpit despite my straps—then to the left as we were flung crabwise into the air again with momentary smoothness. Then another crumpling noise, like ten thousand tin cans being dumped, as she came down on her nose and rocked to her starboard wingtip, slithering sideways and collapsing that undercarriage leg. Another dull boom as the starboard tanks went up. More grinding of metal signified a rapid spin. A stagger as we settled into a dust-spewing heap—then silence, except for the terrifying roar of flaming paraffin. Scared and confused I congratulated myself on still being in one piece; battered and bent perhaps, but not broken.

Two officers had appeared alongside the runway to watch the fun. But my swerve to the right had altered their ideas, and despite the chaotic bumping and bouncing, I was considerably amused to see them run first one way, then the other, like frightened rabbits. With tons of blazing wreckage hurtling uncontrolled towards them, I could appreciate their feelings.

So far so good: now to get out, In order not to disturb the airflow over the broken tail I had left the canopy closed. It had served me well when the tanks went up, saving me from the spray of burning paraffin that followed the rupture of the tanks. But now it was to be my gaoler.

Flames roared fiercely behind, to each side and over me. I could see them, hear the crackling, feel the intense heat. I shut off the fuel cocks to the engines, burning my knuckles on the hot side of the cockpit. I pressed the button which operated the electric motor which, in turn, push backward and open the cumbersome and heavy metal and perspex canopy. But nothing happened. It did not budge.

In a fleeting moment a grim and macabre picture flashed across my mind. I saw again the torso I had found in the ashes of a crashed Harvard. . . . A rectangle in grey-black battledress. . . . There was no head: a three-inch strip of what seemed like leather was all that remained of the neck. . . . Twisted black rosettes represented the shoulder sockets. . . . At the bottom of the rectangle, streaks of darkened blood revealed one thigh joint, while the other was a purple, oozing colliquament of black putrefaction. . . . My gloved hand lifted a charred waistband to disclose human skin. . . . Above everything the smell: the composite odours of flesh and oil, human fat and charred metal. . . . To this day my stomach heaves at the sight of underdone beef or pork. . . .

The picture came, and went. The hood remained jammed closed. The heat was suffocating, and the sides of the cockpit made me wince when I touched them. It had become a devilish furnace. I turned the regulator to "Emergency" to breathe pure oxygen in the acrid, smoke-filled cockpit.

I groped for the crowbar to lever open the canopy, or break the plastic. It was not in its spring clips on the port side of the fuselage. It had shaken itself free, and was later found under the seat. I had no time to find it, and I cursed all designers as the Perspex alongside my head began to melt and sag inwards.

I banged around the cockpit like a man gone mad. I cursed, pressed buttons, pulled, tugged and heaved—but nothing would yield. Neither the jettison handle nor the canopy would give a fraction of an inch.

The Javelin was a blazing wreck, a roaring inferno, devoid of movement. The two R.A.F. officers who had stopped some way off, decided to return to the scene and rescue me. They ran towards the Javelin, but when, with a dull boom, more fuel tanks exploded, and I was covered by another sheet of flame, my rescuers did not pause—they spun on their heels with the precision of ballet dancers, and all I could see of them was two backsides between flailing arms and legs as they rushed from the Javelin. Even then I found it amusing: Lord knows why.

A new thought—I was sitting on a miniature artillery shell capable of hurtling seat and me some sixty feet in the air—through the canopy. What if the heat at the bottom of the cockpit should set it off?

My luck held. At last, persistent attempts at the actuating button combined with banging the canopy with my fist, head and forearms took effect. It shuddered and gave a fraction enough for me to get my fingers under its front arch. The sheer brute strength of desperation helped force the canopy two-thirds open. In a flash I was out and put fifty yards between myself and the blazing wreck.

It was a sad sight. From the broken aeroplane rose a column of black, oily smoke. WD 804 was truly a mess; her great fin was crumpled, her elevators were missing (they were found later in a field near Witney), her nose smashed, her wings shattered, one largely consumed by fire.

People began to collect. As fire tenders rolled up, I dashed back to the 'plane and tried to get at the automatic observer in the nose. But I was driven back. I bellowed at the firemen, and grabbed at their hose to direct their pipe-cleaners of foam on to the nose and tail where the recording instruments were situated. They were the most important things left now; the only bits worth saving. One nozzle gave out, but we got another going, and slowly the flames were brought under control, the instrument bays safe and intact. Now at least we stood a good chance of finding out exactly what had happened.

A flying control type said: "By jove, you're a cool customer, sitting in there waiting for the fire tender to come and put the fire out."

"Do you really think so?" I said. I was in no mood for chat, save to apologize to Air Commodore Macdonald for mucking up the tranquillity of his Sunday afternoon.

An R.A.F. doctor arrived half an hour or more later and did his bit of Empire building by announcing that I was suffering from shock. He popped me into a car and rushed me off to hospital where I was thumped, banged and undressed before my scorched arms were bandaged and my singed eyebrows annointed. I 'phoned the general manager. "I'm sorry, but I've written-off 804—the elevators came off."

"Oh, Lord, what rotten luck. You O.K.?"

After a couple of hours I could stand bed no longer. My foam-soaked clothes were unwearable, so I borrowed others and had a couple of soft drinks at the aerodrome mess. Then the Gloster crew arrived, and I scrounged a lift back to Moreton Valence.

On the Monday there was no Javelin on display for the R.A.F. top brass. Much to everyone's disappointment I made the trip to West Raynham by Rapide. Jimmy Martin, who made the ejector seats, took me to one side and asked: "Why didn't you put your trust in me, pull the blind and take a kick in the backside out?" I replied: "That damned lot of yours is far too frightening."

I believed there were others, too, who would have been happier if I had abandoned the aircraft. Certain boffins and designers never quite forgave me for bringing back the bent 'plane as conclusive evidence. The elevators had gone completely: the victims of "flutter". You can compare it to a barn door swinging wildly in a gale: if it is not restrained, something goes—either the hinges or the wall to which the door is attached.

The records revealed that the teeth-loosening vibrations had lasted no more than two and a half seconds (although they had seemed as many hours), and that the flutter was at a rate far and away beyond those which had been covered in tests. "It's upset all our calculations," moaned the experts.

I further reduced my marginal popularity when a Farnborough boffin was talking about cases of "classical flutter". I interjected, as acidly as I could muster: "For God's sake come down to earth and tell us something about the plain ordinary destructive variety!"

I received a gong: the George Medal, for "saving the records after a crash landing". They had to say something, I suppose. Our publicity boys earned their keep by building me into a tin-pot hero, the argument from their point of view being a sound one: build up the pilot as a man of the moment, to divert critical thoughts and press reports from the aeroplane.

14

"YOU'VE GOT IT"

After the Javelin prang, I felt that I could not quit immediately. People would think the crash had "finished" me, and I did not like to leave a job half-done.

By a strange freak of fate, John Derry of de Havilland's came to Gloster's job-seeking that summer. The idea was that he should do six months or a year as my number one, then take over. Negotiations broke down, and John continued with his old firm.

A second Javelin prototype came along just before Farnborough, and I flew it at the show. De Havilland's 110, a far less revolutionary 'plane, piloted by John, was our closest rival, and on the day before the show ended, John "bought it". The 110 fell to bits in the air, killing John and Anthony Richards—his observer—and caused the death of twenty-eight spectators, injuring a further sixty-three.

I stood beside John's wife, Eve, as it happened, and for a few minutes afterwards. Then it was my turn to go up. . . .

The controls of the second Javelin were, if anything, worse than on number one; especially the elevators, which had extra weights on their noses in the hope of preventing a recurrence of "flutter".

I was still looking for an experimental pilot to serve as my number one and eventually take over, but it was no easy task. There were few men prepared to do the work for the low pay I was permitted to offer. It had to be someone who was in love with aeroplanes for their own sake—and I finally discovered such a person in Peter Lawrence, an ex-Naval pilot who had been with Blackburns. He joined us in the late autumn of '52, and his appointment did not go down at all well with some of the other pilots.

They had wind of his coming, and promptly asked me about it—particularly two pilots who fancied themselves as experimental flyers. The more eloquent was a man who'd had his break when I put him in charge of training some Brazilian pilots. Among his duties was the job of selecting and buying furniture to furnish a room for them. This he failed to do. In fact he disappeared into thin air, to reappear only when the Brazilians had actually arrived. In the meantime I'd done the furnishing myself, and had also laid out a training programme—which he should have done. His chance had come—and gone.

Finding office space and quarters for the Brazilians had been another incidental headache. No one was prepared to give up space or put themselves out, even though the Brazilian order meant an assured six months' work for the works: the result of a brilliant and unexpected deal by the company secretary whereby Gloster's swapped 'planes for cotton. In the end I moved out one of my pilots who, in my absence abroad, had grabbed an office and secretary, and put his name upon the door.

By 1953 production had slowed considerably; fortunately three pilots had already left the firm. Work would have virtually stopped altogether had it not been for the Brazilian order, and additional orders from Israel, Egypt, Syria and Denmark. These kept us quietly occupied.

There were still five pilots, and ferrying aircraft out to the Middle East was about all there remained for the production boys to do. Some time before, the government had cut back Meteor orders by some two hundred aeroplanes, which meant there would be a big production gap between that model and the new Javelin.

The ferrying provided me with another headache. The pilots were on a "good thing", but even so the petty squabbling and bothers that came to my ears made me envious of theatrical agents who had only prima donnas with whom to deal.

As the Ministries of Civil Aviation and Supply got into their strides so did the paper work pile up. From a trickle in 1947, the red tape and regulations had become a torrent as bureaucracy snowballed. I could easily have devoted all my time to paper work without ever looking at an aeroplane. The only way I could begin to keep-up was by night and weekend work, for it was a lone task. Within the firm the same thing had occurred. With decisions and backing difficult to get, frustration was complete. Altogether, these bothers cost me many a night's sleep—sleep that could ill afford to be lost with a new 'plane and tricky trials on one's plate.

The second Javelin prototype was sent to the works to have a set of new-style wings. These had their leading edge sweep reduced from about half span to the tip, the idea being to aerodynamically thin down the outer wing, "clean up" the ailerons, and reduce tip stalling which causes aeroplanes to automatically tighten into turns. The former wing form had fallen down badly on useable high Mach No., and was not producing the required turning ability at high altitude.

While she was away, the third prototype, WT 827, came along, and I first flew her on March 7, 1953. When the second one, WD 808, was returned at the end of May, I made the first twelve or fourteen flights to establish her characteristics with the altered wings, then turned it over to Peter Lawrence for a second opinion.

I was not altogether happy about the 'plane at low speed, for elevator control was sluggish, without much feel. But Peter had flown the old type for some hours and was familiar with the aircraft. He had been briefed on handling with the new wing and had studied my reports about it. So I felt quite confident in letting him go up in her.

On the morning of his flight I was at my office, some twelve miles by road from Moreton Valence, having left instructions that after one trip in the Javelin Peter and I would talk things over. But when I returned to Moreton Valence after lunch I found Peter in the cockpit ready to start up again—on the firm's instructions, not mine. I discussed the morning's flight with Peter, and since it had apparently gone smoothly and he was an experienced experimental pilot who would only be covering old ground, I did not pull him from the cockpit and raise merry hell with all responsible. I wish I had. . . .

Some half-hour later the flying controller rang up to tell me, in great agitation, that a message had come through from the Javelin to say it was in difficulties. . . .

I rushed across to flying control. Over his R/T, Peter had called: "I'm in trouble. . . ." There were a few more garbled words—then silence. Repeated calls from the tower failed to raise him. As part of our emergency routine we called other aerodromes. Vickers at Chilbolton had heard much the same as we, and their automatic direction-finding equipment placed the Javelin at the time of its last call on a line running north-westerly from them and just south of Bristol. There was nothing we could do but wait. . . .

The tense, silent agony was soon broken by a call from the Somerset police. A 'plane had crashed near Bristol. . . . Something had come out of it. . . . It had just missed a crowd of boys playing cricket. . . . The pilot had been found in his ejector seat some way from the burnt-out wreckage. . . . He was dead. . . .

I jumped into a Meteor, located the site of the wreckage, and confirmed that it was the Javelin. Back at Moreton Valence I informed the general manager, then set off with my secretary, a nurse and my flying control girl to break the news to Barbara, Peter's wife. I had only met her a couple of times, and hardly knew her, but, as always, it was left to me to break the news. It was, I felt, vital to let her know of the crash before she heard of her husband's death from some other source, for even if the B.B.C. mentioned a "crashed Javelin" without naming the pilot, she was aware that her husband was one of only two men who flew it.

I found Mrs. Lawrence preparing dinner and performed my unpleasant duty. She stood up magnificently to the tragedy, and I stayed with her until the evening, when her relatives arrived.

From what we were able to piece together, it appeared that the Javelin had got into some sort of stabilized stall—had dropped almost vertically like a brick at more than a mile a minute, with very little forward speed. Such records as were salvaged indicated that Peter had come down to a speed where warning of the stall was given, recovered, then approached it a second time, with flaps down. It then seemed to have rapidly and suddenly got into a condition from which he did not regain control of the 'plane, although he fought it all the way down, only ejecting himself at the last possible moment. But he was too low to get clear of his seat and use his parachute.

Why the trimming tailplane was found to be the wrong way for recovery was never established, but it may have been significant that the 'plane was the first Javelin to have a trim button fitted to the stick instead of the trim wheel fitted to the first two Javelins. I knew from experience that this button could be misleading, for on an earlier flight trim had gone the wrong way to which I intended, and there were other disturbing incidents with other aircraft.

As a result of the accident the firm altered the flaps. It was little consolation to me that it took the death of my number one to effect modifications that my talking and reports had been unable to achieve since the aeroplane's earliest flights.

Nothing was disclosed by the inquest: it rarely is when new planes crash. Coroners do not ask compromising questions of firms engaged on secret work. This has often struck me as being unsatisfactory, for what chance have next-of-kin of taking action against any possible guilty parties when everything is cloaked in secrecy? Results of an investigation are never submitted or released, and I never saw the official report or findings concerned with this disaster—although Accident Branch investigators were on the spot and I asked Air Ministry representatives for a copy. In the Services, as well as in civil flying, all the information relating to accidents is not made known. I can think of no parallel evasion of the law—a cause of death (or its contributory factors) being kept hush-hush.

I again had a severely restricted Javelin when the 1953 Farnborough came along: it seemed as though we were destined never to have more than one prototype going at a time. During the winter 1953/4 the third prototype had her guns fitted, and the gunnery trials went well, despite a few stoppages due to belts fouling, and bother with gas concentrations. (Sides had been blown out of Meteors due to explosions in the gun bays: gases seeped back from the breeches of the firing cannons, and under certain conditions, if concentrations were sufficiently heavy, they could detonate with disastrous results. The boffins fitted meters to record the gas concentrations, and when the needles rose above the danger mark one had to cease fire quickly—and keep one's fingers crossed.)

Previously I had carried out "shaking trials" in this aeroplane, acting as a flutter "guinea pig". Flutter was plaguing many British 'planes at the time—not only Gloster's. So the second prototype was elaborately instrumented, and in her nose were eccentrically mounted weights which were operated by electric motors and spun to shake the 'plane in flight. Hour after hour I flew at varying heights and speeds, applying a large range of vibrations, and hoping no part of the 'plane would "get into step" with any of them—like a glass breaking through resonance when it is in tune with a violin.

Then the automatic ejection seat was fitted to the Javelin. When the pilot pulled the "blind" over his face it automatically took over. At 10,000 feet (or after five seconds if below that height), it would kick the pilot from his seat and open his parachute. But such was the logic of designers that the pilot still needed to find the cockpit canopy release and jettison the canopy before ejecting himself. It was not then possible for a pilot, in a moment of near panic, to perform both operations in one movement.

Cases had been known of pilots ejecting themselves *through* the plastic canopy. On those Javelins this was hardly feasible; strengthening metal bars had been placed above the pilot's head! One thing was certain, anyway—the boffins and designers were never likely to use their creation themselves in an emergency.

There were other features and alterations which in the circumstances were no less disconcerting. These concerned certain cockpit arrangements and dimensions which the fitting of newer equipment had altered. As usual, the pilot's job was made far more difficult by the changes and he was less comfortable than previously. The D.O. staff, when informed of what had transpired, expressed a certain amount of hurt surprise but no real concern.

During these trials another old trouble came back with a bang—literally with a bang.

There had been trouble with our previous types of aircraft, which were reluctant to keep undercarriage doors fully closed at all times, and when I first worked up the Javelin to her full design speed low down—early in 1952—I came in to land with the port undercarriage doors torn away and gone. Then, two years later, returning to Boscombe from the sea ranges at Lympne during gun-firing trials, the same thing happened. At 600 m.p.h. there was a sharp crack, rather like a cannon being fired, and the 'plane lifted and shook as though an ack-ack shell had exploded under her. Another door had let go—again on the port wheel.

A time-consuming job was the sorting out of electrical equipment. This was always a nuisance on our aircraft, and a constant source of trouble. (Half the time prototypes spend grounded can be accounted for by ancillary problems, since everything in a new 'plane is specially designed and constructed for it. I have always felt that much of the equipment need not be made-to-measure, and that the saving of a great deal of time and public money could result from the standardization of certain parts.)

A modern, all-weather fighter makes enormous electrical demands to run its radar, radio and navigational devices. This power generation—and its conversion into D.C. or A.C., generates heat: heat which will cause the equipment to fail—even destroy it—if it is not dissipated. Paradoxically, at high altitudes, where the temperature is often a hundred degrees below zero, cooling difficulties are at their worst: the air is thin, its particles far apart, and heat absorption is therefore small.

So, with special measuring instruments read by another pilot in the back seat. I would fly at high and low speeds, at various rates of climb, from just above ground level to 50,000 feet. Up there pressure waistcoats and pressure breathing masks were essential to squeeze oxygen into the lungs in case the pressurized cockpit failed. On occasions it did fail—and it was positively eerie when the warning horn went.

Although I went beyond a Meteor's best, I never took the Javelin to her maximum ceiling, for although I had a pressure suit that would allow one to work at really high altitudes, Gloster's never had a 'plane modified to take it during my time.

Adapted from an American design, the Service-type pressure suit is a zipped and laced fabric affair worn close to the skin. It is topped by a pressurized helmet through which sounds and radio noises percolate with muffled remoteness. When "On", pneumatic tubes tighten the fabric, putting on the body a pressure over its entire skin surface. If the fit is not perfect, the body often comes out a mosaic-work of bruises from damaged blood vessels. Nor is the equipment automatic. If cabin pressure fails, the pilot has to switch the flaccid suit over to "Pressure"—then nip downstairs.

When the Javelin came along, so did the Americans. In the early days of the Meteor, we had been prepared to fight them commercially. Now the opposite applied, and the management made determined efforts to woo our trans-Atlantic allies. The Yanks, always excellent business men, were not prepared to throw their money away on an unknown product, so their Air Force representatives came to Gloster's to fly the 'plane. This meant that they would be flying the Javelin before our own R.A.F. boys, and I thought it a bit thick. I set wheels a-turning, and four R.A.F. pilots attached to the M.O.S. moved in smartly from Boscombe Down to fly the second prototype. Then the Americans arrived, headed by General Al Boyd. But the weather was against them, and they went away without flying.

Eventually, in the spring of 1953, Colonels Dick Johnson and Pete Everest flew the second Javelin, bringing Sabres along as pace setters. It was a tremendous satisfaction to me that their impressions of the 'plane and its defects coincided exactly with my own. The faces of the designers were a study as they listened to the Americans' criticisms: it was obvious that they had not believed a word I had written in my reports, but here were two impartial airmen repeating virtually everything I'd said. Yet, incredibly, when the Americans left, their complaints were airily dismissed for the most part, and things continued as before. The Ministry of Supply receive six copies of all flight test reports, and their representative sat in when the Americans gave their views. It would be interesting to know what that Government department thought of it all.

A sidelight on inter-departmental rivalries is revealed by my efforts to get an R.A.F. pilot from the Central Fighter Establishment to fly the Javelin. Time after time I nagged, verbally and in writing, for permission for the Night Fighting Squadron's Wing Commander and Squadron Leader at C.F.E. to fly the 'plane. They were not test pilots in our sense of the word, but they were the chaps who could assess the Javelin as a useable *fighting* machine—which, after all, was what it was supposed to be—as distinct from the M.O.S. pilots at Boscombe Down who were primarily concerned with it as a *flying* machine.

But the M.O.S. said no, until ages later when a very high-level wangle brought Wing Commander Hughes to Moreton Valence. He only had the afternoon free, and as the weather closed in, he was able to do little more than just go up and down in the 'plane. It was plain that the Ministry at that period did not want Fighter Command to get their hands on a new 'plane until Boscombe said so, and by that time the aircraft would be in production. Then, and only then, could the R.A.F. assess its true value as a fighting machine. Since those days, I am pleased to say, the set-up has been altered.

Before the U.S.A.F. visit I had explored the 'plane's speed range at low and high altitudes and, according to the boffins, had pushed the Javelin through the speed of sound. The firm was naturally delighted when a report was sent to the Ministry informing them of it, but such flights were no pieces of cake.

On one occasion I put the Javelin's nose down from a great height and really got moving. After the normal phenomena—lateral shaking, change of trim, pitching and buffeting—I became aware of something else. A deafening row, like a battery of machine guns, started in the back end. The rudder bar kicked from side to side, and the 'plane weaved violently as she hurtled downwards in a tran-sonic dive.

Although we kept quiet about it, the newspapers reported "heavy explosions in the Bristol area" at the time my tail-wagging dive was in progress.

The production Javelin was to be virtually the same as the fourth prototype. I have vivid memories of my first flight in her on February 28, 1954: it was so typical of the way things went.

To help out the works, I said I would fly the 'plane direct from Brockworth, where it was made. There were only 1,000 yards of concrete runway, and although previous flying had shown this to be enough, I was still a little apprehensive, for there had been trouble with this 'plane's hydraulics, and I wasn't too happy about the new ailerons; if anything went wrong, I could not abandon take-off and stop short of the trees at the aerodrome boundary. Once 250 yards had been covered I was committed.

WT 830, as she was numbered, was pushed out on the tarmac near my office. I ran the engines, and she was then towed along the narrow perimeter track to the runway, where she was refuelled and prepared for flight. It was late in the day, yet it was highly desirable from the firm's point of view to say that the Javelin should take the air, for it sounded better that we flew our fourth prototype in February (albeit the 28th) than in March (even the 1st).

At last I was told that all was ready. I left my office, parachute over my shoulder, and was a hundred yards or so down the tarmac when the junior secretary came tearing after me. "Oh, Mr. Waterton—telephone."

"Hell, I can't bother with it now," I said.

"It's very important, Sir. It's Mr. So-and-so."

"In that case," I conceded, "I'd better hear what our technical office has to say."

"Hello, Bill," said the voice at the other end of the 'phone. "You'll hold off for a bit, won't you ?"

"Sorry, but I can't. It's getting dark. What's the trouble anyway ?"

His voice was hesitant and apologetic. "Well . . . We're not too sure about those new ailerons. . . . We've just discovered they're understressed, and we'll have to put a limit on the machine."

"Well, what *is* the limit ?"

"Er . . . er . . . we don't know yet."

I tried hard to contain my impatience. "I see. What does the limit look like? Three hundred? Four hundred miles an hour?"

"Heavens, no ! It's more like two hundred !"

"Good God—and the 'plane's already three months' late. . . . All right. I can hold it down to well under that if needs be. All I want is enough to get her to Moreton Valence. . . ."

I took off, navigation lights glowing in the gathering dusk. The first production Javelin, with full power ailerons in the new type wing, had left her factory aerodrome—yet not one director was present to see her off. I put the Javelin down at Moreton Valence without incident, and she stayed down—for aileron modifications.

When this sort of thing happens, it hardly strengthens one's confidence. After all, there are enough things to go astray without the test pilot having to worry about the integrity of a 'plane's engineering.

When, for example, you handle power controls which you can hardly shift with two hands on the ground, with no air load applied, surely you are entitled to ask : "Is this good enough ? Can't we have a trim tab on the aileron ?" But you don't expect to be told in reply: "It isn't necessary. It won't go wrong." Such an answer is not good enough—especially when it was this very thing that eventually did go wrong.

When you know that things are not right for you—far less for junior squadron pilots—it is more than infuriating to be told by a member of the management: "Pipe down. You're too critical. We've got to get some of these out. . . . It'll be all right."

When senior R.A.F. officers at the M.O.S. suggest that you should not be too critical because they want the aircraft in service in a hurry, it's time to pack up if you've got a conscience.

Yet that was the position in the spring of 1954.

Meteor production, down to a trickle, was scheduled to end in March. Our three production pilots were asked in turn to go to Brazil to look after the Meteors delivered there, but each declined. So we took on an ex-Rolls-Royce pilot who, getting on, had been generously pensioned off by that concern.

With production at a standstill I was told by the management to sack two of the production pilots. I retained, in the firm's best interest, the youngest and keenest. At Christmas I had taken on, as an experimental pilot, ex-Cranwell cadet and former Farnborough test pilot "Dicky" Martin.

I was disgusted at the way the firm treated the older of the two sacked pilots, for he had done a lot of good production testing. It is true that they gave him a few months' pay, but it was an insulting pittance when compared to what other firms did for their old pilots. No, ours was not a firm in whose service a pilot could happily grow old. I recalled a certain director's charming smile of seven and a half years' earlier: "Oh, Bill, I do hope you will come and work for us"; and the words of the management when the question of my Permanent Commission in the R.A.F. came up: "Don't worry, Bill, we'll see you all right. You can depend on us. We'll look after you." As my old aunt used to say: "Talk is cheap."

By March 11 my mind was made up. I went to the general manager, and said: "Where do I go from here, Sir ? I really must know."

"I'm afraid I can't tell you," he replied.

"Have you any objection to my seeing someone who can ?"

"No, Bill, go ahead."

So I arranged an interview with a very senior director of the Group; a man who was associated with a score of companies and was a great power behind the Gloster throne. A man, incidentally, who had said a couple of years before: "I hear you're thinking of packing up, Bill. Don't worry, I'll sort out something good for you." A man who prided himself on his fairness.

He greeted me cordially enough, and was radiating affability. Yes, he said, he understood the situation.

I told him that I wanted to know what my future was going to be. I said that I felt badly underpaid compared to the chief test pilots of other firms who had but a fraction of my experience and responsibility. I wanted some assurance that action would be taken to put right my many criticisms of the Javelin, and that I would have returned to me the authority that had been taken away while I was overseas.

He agreed with all I said.

I added that I simply wasn't prepared to go on with things unless I got some concrete satisfaction. Otherwise, quite simply, I would be forced to resign.

"See me in a week's time," he replied. I went away, leaving him with a written report of what I had accomplished while at Gloster's: an account of the near-3,000 aircraft which had been through my hands, how I had at times run three aerodromes—the whole story, written in concise, statistical form. I concluded by repeating that if my complaints were not met I would have no option but to resign allowing three months in which to hand over to my successor.

The Rubicon crossed, I felt much better, as is always the case when a worry, doubts and uncertainty have been replaced by positive action. Whatever happened now, my cards had been put fairly and squarely on the table.

When I returned to his office a week later, he said that we'd go to see my general manager. He had his car, and I mine, so we made our own way to the other's office. When I arrived, however, the boss was alone. He seemed fidgety and ill at ease. I knew what was coming before he brought himself to blurt out: "With the frame of mind you seem to be in, we think it best that you go."

"Right," I said. "When ?"

Then came a bombshell: "Tomorrow."

"But that will hardly give me time to hand over the job. . . . In any case my number one isn't ready for an immediate transfer of duties."

"That's our problem. . . . It's better to wind these things up quickly."

I reminded him that seven and a half years' correspondence and equipment couldn't be sorted out in just one day. I pointed out that there were still many immediate detailed troubles with the Javelin that needed putting right, and that the fourth prototype, which had suffered a hydraulic control failure in flight with an R.A.F. pilot, was being patched up at Boscombe Down, and needed bringing home. In any case, I said, I can't push a new 'plane into another pilot's hands without a thorough briefing. "Unless you forcibly chuck me out, I'll not leave until the end of the month, in eleven days' time."

"You'll always be welcome here," he murmured.

The senior gentleman had arrived meanwhile, to back-up the other and also, no doubt, as a witness. He announced that I would receive some compensation and mentioned that I could buy a car cheaply from the firm. I said: "Thank you very much, but I am already driving my own car," and expressed regret that I had been so foolish as to devote seven wholehearted years to the responsibility of his aircraft and proto-types. There being nothing more worth saying, I departed.

I continued with my job as though nothing were amiss, quietly gathered together my correspondence and belongings and went on flying as usual. I brought the Javelin back from Bo\scombe, and briefed my number one on it. The Ministry of Supply showed no apparent concern that almost all the handling and flying background to their important aeroplane was to be lost immediately, and I was never to have any further similar contact with that body which had helped rule my working life for so many years.

Rumours circulated, but nothing definite was known until March 31, when I went round making my farewells. It came as a bombshell to most people, and many weren't actors enough to conceal their pleasure at my departure. One said: "Shall I say I'm sorry?" I replied: "Hell, no, why perjure your soul at this stage of the game ?"

I flew my last aeroplane that afternoon, tested one of the last production Meteor VIII's, had a photograph taken to commemorate the event and drove out the gate of Moreton Valence for the last time. I had done exactly seven years as Gloster's chief test pilot. Tomorrow was another April Fools' Day.

WHY BRITAIN HAS FAILED

What I have to say here is not directed against any individual or firm: it is intended as an overall indictment. For a parlous state of affairs exists *throughout almost the entire airframe industry*, and members of the Society of British Aircraft Constructors (together with Government officials, Services chiefs and civil servants) must share the burden of responsibility.

An individual firm is only publicly limelighted when a particular project, after enthusiastic advance publicity, is proved a failure. But virtually every firm has its unsung, discreetly hidden mistakes.

Many people knew, for example, that the Bristol Brabazon was an acknowledged flop before it was half-completed. Money spent: a reputed twelve million pounds. There was the great Saunders-Roe Princess flying-boat let down by its engines, and written off for its original purpose at an estimated ten million pounds. A further twenty were said to have been spent on the Supermarine Swift. It was hailed as a world record-beater, issued to the squadrons—then withdrawn as a failure. Now it has been salvaged to appear in the rôle of a fighter reconnaissance aircraft.

But there have been others, to swell to even more gigantic proportions this figure of £42,000,000—almost all of it public money.

Yet no major aircraft company has closed down since the war, irrespective of colossally expensive failures. Indeed, they would not be permitted to, for two reasons: politically it would be unsound to throw thousands of people out of work, and it would be strategically unwise to allow a firm to put up the shutters when, in a national emergency, it would need time to take them down again. And firms know this.

Illustrating this is the case of the post-war fighter which neither the R.A.F. nor the Navy wanted. But it was built in quantity nevertheless because (and the story is an open secret) the manufacturing firm told the Ministry of Supply: "Either we get an order or we close down." Blackmail ? An ugly word. . . .

Nor is it easy, when an aircraft flops, for one man to be accused as the guilty party. He is only one cog in a gargantuan, creaking machine.

It all starts when the requirements for a new 'plane are drawn up by the Service or airline concerned. Since five to seven years will pass before the 'plane gets into service, considerable crystal-ball gazing is inevitable. Needs are largely determined by (a) what the "other chap" is likely to put in the air at that time, and (b) what is possible technically and what manufacturers say they can do. Invariably (b) decides the day, irrespective of requirements or anything else.

Yet the industry is often defeatest in its estimation of what can be achieved technically—not surprising when it has failed to exploit the latest in tools, techniques, materials and ideas. I remember the R.A.F. asking for a clear-vision cockpit canopy, only to be told it was impossible. None the less, American Sabres were flying at the time with just such canopies—not the vision-restricting hoods of British fighters with their great area of metal. So fed-up was the R.A.F. about this that Central Fighter Establishment got their hands on a couple of Sabres, took a canopy from one, went to a contracting firm and set it on its fighter—just to show that it could be done.

This is no isolated case. Time and again I have known the R.A.F. and M.O.S. to be told they could not have what they wanted—and they seemed powerless to do anything about it. (Subsidized by the Government, the aircraft companies are on a safe thing: whoever loses, they win. They sit tight—and smug.) Emasculated by safe Government contracts, none of our manufacturers has had the courage to invest his money in a much-needed light aircraft. In the same way, we have no helicopter to compare to the Americans', and no proven long-range civil airliner (with the exception of the Viscount and possibly the Britannia).

I digress. . . .

When the customer has decided upon his needs, an official specification is issued to approved firms by the M.O.S., and those interested submit design studies from which usually two are chosen. They might be radically different from each other, as were de Havilland's 110 and the Javelin, and the Vulcan and Victor, or remarkably similar, like the Swift and Hunter. For insurance reasons (and to keep the industry busy) both firms are set to build prototypes, and orders go out for ancillary equipment. (Here, as I have said, there is a strong argument for standardization: time and money could well be saved if a strong directive urged—and challenged—firms to wrap their shapes and new ideas round common wheels, brakes, generators, etc.—as they do engines and armament.)

At this stage, and throughout, payment is made for design work, materials, tools and tooling, jigs, development work, flying, modifications and changes. An order is guaranteed for production, and to the lot is added overheads—often a hundred per cent.—plus a fixed profit. This is known as "cost plus" and the more the cost the more the plus. Tools and buildings are loaned or rented to firms and if contracts are slashed or 'planes unsuitable the firm is paid compensation.

Within three to six months of its first flight, the general pattern of the prototype's behaviour and performance is usually determined. This is something that cannot be rushed, for although the customer ought to come into the picture early on, a firm must be granted a reasonable period in which to make necessary modifications: a project that starts badly might work out well—and *vice versa*. But no more than a year should be needed, and firms made to work to that deadline. At the end of those twelve months there is no reason why one of the two prototypes could not be selected—although not by examining the results and figures presented by the manufacturers, as often happens now. Instead, it should be done as we did it at Central Fighter Establishment—by the practical method of flying one 'plane against the other in side-by-side climbs, accelerations, decelerations, dives, tail-chasing turns and rolls, with camera guns firing. After such trials there would be no doubt of comparative performances, for even mock attacks are a thousand per cent. more reliable than paper figures and individual tests. Yet, incredibly, these vital and logical trials do not come until a 'plane is actually in production.

Shortage of prototypes is another time-wasting bugbear, for if you lose one or two very special aeroplanes, as we did with the Javelin, progress is delayed for months—even years. Recently Air Chief Marshal Sir John Baker, the C.A. (Controller Air) pointed out that twenty English Electric P.1's had been ordered to speed development. Had the new prototypes come along at regular, frequent intervals of, say, three months in the first place, it would have been something to shout about, but the second did not arrive until about a year after the first—the same as in the past.

Once the new aircraft has been selected, the other should be dropped without more ado—unless it has qualities to suit it for some special rôle. Both firms should then concentrate on *producing* the new 'plane; the winner's design staff dealing with technical problems and changes as they arise, the loser's getting to work on fresh designs for the future. As things stand, only the winning firm produces the new 'plane, while the other ambles along often manufacturing old stuff contracted to keep the workshops occupied. Otherwise, both are given orders for their separate 'planes resulting in double sets of costly jigs, tools, ancillary equipment and testing for minute production quantities. This is presently happening in the case of the Victor and Vulcan, making for high costs per production unit and duplication headaches in R.A.F. stores, ground equipment and training, both flying and technical.

Let the design staff admit their faults, and if too many occur, break them up and install people who are competent. Faults are common to all new aircraft, and are nothing to be ashamed of. Let there be an end to this business of "getting by", ignoring what the test pilots and ground servicing people say, and covering up. It should not be necessary to wait until someone is killed, or until faults are spotlighted in service and 'planes grounded *en masse*, before modifications are made.

The trouble is that few British firms understand development work. A new prototype is built—and that is pretty well that. Consequently our production aircraft do not fly at all as well as they should, and are rarely little changed from their first prototypes. The users get 50 per cent. aeroplanes instead of 90 per cent. aeroplanes. We could learn here from the Americans. They ran into serious trouble with their Super-Sabre, and their Convair Delta F102 was badly down in performance. Yet within three months the Sabre was comprehensively altered—given a redesigned tail, controls and wingtips—and was out of its troubles. The 102's faults were corrected with equal hustle. Britain has demonstrated nothing to compare with these methods. Witness the Comet, for example: a brilliant conception, let down by its aerodynamics, engineering and handling—nothing like a 100 per cent. aeroplane. Externally, the Javelin, Hunter or 110 have hardly altered since prototype days. There has been no wasp waisting to make them conform to the area rule and so raise speeds by up to 25 per cent.

Under existing arrangements, the people who design the 'planes are usually responsible for their development and, like proud parents who have produced a misfit, they are reluctant to admit the fact, and are furious when other people criticize. As I see it, when a prototype flies it should be taken right out of the hands of the designers (who thereafter become no more than consultants) and passed to fresh minds, dedicated to making the 'plane efficient as quickly as possible, regardless of all other considerations.

The Services blame the M.O.S. when the right aircraft do not arrive in the required numbers at the proper time. It is true that the Ministry has much to answer for, but the Services cannot claim not to know what is going on. Both the Navy and Air Force have officers attached to the Ministry, and an airman is Controller Air. He is responsible for ordering, and for controlling testing and development, and since he has a seat on the Air Council, that body can hardly plead ignorance of the state of the new aircraft and their faults. The R.A.F. and Navy may not be getting the aircraft they want—but they seem to be keeping pretty quiet about it.

These are some of the factors contributing to the overall picture of the muddle, inefficiency and lethargy which are in varying degrees responsible for Britain being almost an also-ran in the aircraft stakes. If it is doubted that we do only just manage to scrape into third place—trailing behind America and Russia—consider how their development has leapt ahead. Both have produced in quantity fighters which can "break the sound barrier" in level flight, and heavy bombers are in service twice as big as our largest. Soon a United States' bomber the size of our V-class machines is to be flown at supersonic speed in level flight, and the Americans have flown 500 m.p.h. faster than any Briton, and a good deal higher. The Americans claim, further, that they have four fighter aircraft capable of winning back any new record our P.1 could set up, and knowing a considerable amount of both sides' claims, I do not doubt the United States' boast. We have dropped flying-boats while the Americans have progressed with advanced designs, and there is the lack of helicopters and light 'planes to which I have referred.

With safe Government contracts, our manufacturers lack the incentive of real private enterprise to challenge the Americans and Russians. In all but name and the distribution of profits, they are already nationalized in a way. Nor is there the incentive of pride—the pride of airmen—for the heads of the industry are almost exclusively financiers, accountants and business men. (One notable exception is Rolls-Royce, where the executives are engineers first and administrators second.) Experience has led me to believe that heads of firms fear the return of a Labour government and the threat of nationalization, and so argue, "The Socialists will have the lot so let's grab what we can while the going is good." They have further covered themselves by pouring money into overseas plants. And remember—an aeroplane factory is equipped to manufacture many articles, so the change-over can cope with a variety of circumstances, especially overseas.

One thing is certain: the firms have not ploughed back the money they should have done. A walk through a British aircraft factory and then an American or Canadian one would soon prove this point. By comparison our firms are back-alley garages. Even though some of our groups and enterprises boast of over 60,000 employees, they are composed of a mass of small units, more often than not working against each other or duplicating each others' efforts. There is not one firm in Britain which could manufacture 'planes of the size of the defunct Brabazon in quantity. What firm here has the plant or tools to build the one hundred-plus giant airliners ordered from Douglas? They lack the vast presses, stretch presses, milling machines, shapers, drop hammers, and even the abundance of small hand-power tools of North America, and as a result we are building 'planes almost identically in the way we did fifteen or twenty years ago, despite the revolutionary demands of the jet age. Javelins are built in much the same way as Spitfires, and there are none of the heavy rolled or milled "skins" used in America, and only a token use of titanium. And, this delay of the airframe structural revolution hinders and limits aerodynamacists and designers.

This modernizing of our factories is a priority task, for as things stand we cannot introduce even existing American designs—far less think of progressing ahead: we haven't the means of transferring them to the production belt.

Not only have we failed to keep pace on the engineering side, but we are way behind on the aerodynamics which dictate the shape of new aeroplanes. For years few companies, for instance, had their own wind tunnels. Farnborough did most of this work and, not unnaturally, was overloaded, with the result that many tests were left undone. High speed and supersonic tunnels are still at a premium. The lack of these tunnels has meant the absence of much important research, and we have tried to muddle through by guess and by God. Logically, such methods are impractical in the jet age. When the United States sent her pilots through the sound barrier for the first time, the flyers knew, from ground missile and wind tunnel tests, what to expect. Our chaps still have to "suck it and see" when exploring new ground.

The Government has been blamed for our lack of full-scale research facilities, and although it is true that they have passively done nothing to shake things up, it must not be forgotten that the industry, operating on public money, has made vast profits in the past ten years, and insufficient of it has been ploughed back for this purpose.

So we see that in both research and engineering facilities we are way behind current requirements, and there is yet another factor to consider: personnel.

There are keen brains and excellent engineers and aerodynamicists in the aircraft industry. There are also many deadbeats—a hangover from the war and pre-war years; people, many in responsible positions, who are hopelessly out of their depths, and who are doing their damnedest to see that no one who knows his stuff is likely to reach a position where their shortcomings will be laid bare. They exist at all levels, from director to labourer, and they haven't done a decent day's work for years. With many it is politics, first, last and always—not "is this the best way to do the job; will this produce the best possible aeroplane quickly and cheaply?" but "how is it going to affect me and how much can we sting the Ministry?"

So the good men are kept down—even forced out—by the bad. Pay, too, is generally far from generous. Only recently an employer said to me: "We're trying desperately to get aerodynamicists, but they've got the nerve to want a thousand a year." During the war the industry was able to get all the brains it wanted, and cheaply; today the mathematicians go elsewhere—to football pools firms, for example. Even a chief aerodynamicist, the man who determines, lays out and advises on the shapes and sizes of aircraft and their parts, often receives little more than £1,500 a year. Ten thousand would not be overpayment for a first-class man. To my mind this is one of our biggest failings. Directors baulk at the thought of any one individual under them getting big money. They revolt at paying two competent experts £50 each per week, yet cheerfully pay ten incompetents £15 to £20 per week to muddle along and accomplish nothing.

There, then, are the main reasons for Britain's failure: the smugness of firms whose initiative has been destroyed by safe Government contracts. . . . Dilatory and inefficient methods and the lack of proper organization. . . . A failure to understand development work. . . . Lethargy on the part of the R.A.F. and Ministry of Supply. . . . The shortage of engineering and research facilities. . . . The choking effect of lay-abouts and hangers-on. . . . A general tight-fistedness in the wrong directions which, among other things, prevents the industry from obtaining, and retaining, the best brains available. Last and most important is the failure at all levels to think and act big.

How is the situation to be remedied ? As things stand no one at a sufficiently high level anywhere has had guts enough to stand up and call the cards. No Service chief has yet risked his rank by revealing the truth. Nor has any M.O.S. official. One or two M.P.s often hit the nail on the head, but the situation demands far more than lone voices from the Opposition back bench.

I feel that nothing less than a Royal Commission will do to investigate thoroughly the aircraft industry and the procurement of aircraft— one whose findings will not be hidden by dust and quietly forgotten, but a body whose conclusions will be acted upon without delay. For the sands are running out.

The aircraft industry, the M.O.S., the Services, air transport firms, airlines, all need looking into. Indeed, so does the nation's whole aviation policy, for there are too many sectarian interests at work in divergent ways. A strong man is required, for only by ruthless measures will things be changed. If the Services do not get what they want they must say so—and the responsibility laid fairly and squarely at someone's door. Contracts for specifications, price and delivery must be honoured. If a firm fails, let it fail and be taken over as a national arsenal. The industry talks private enterprise; very well, let it take the risks of private enterprise as well as the profits.

There is nothing wrong with British air matters that honesty, frankness, ruthlessness in the right quarters, and hard work, cannot put right; but it must start at the very top, or a lead must be given from the very top. The well-being of the entire nation is above that of individuals and firms.

Gloucestershire and London, 1955.